Cultural Adaptation

Cultural borrowing is exploding across the world. Creative ideas are transferred and modified in ever increasing number and complexity making new products ranging from TV shows to architectural style in new cities. But what do we really know about the spread of creative ideas? This intriguing, engrossing, and comprehensive collection looks at the cultural and commercial dimensions of creative borrowing world wide with an international cast of contributors and case studies from India to Ireland, Canada to China.

Cultural Adaptation explores how creative ideas are packaged and nationalised to meet local taste, maps the cultural economy of adaptation in entertainment media ranging from motion pictures to mobile phones, and even probes the role of cultural recipes and formats in mutating participatory experiences of theme parks and sporting spectacles. Written in a lively and accessible manner, the book also provides insight into remaking in lifestyle and consumption cultures including fashion, food, drink, and gambling. Essential for communication, cultural, media, leisure and consumption studies scholars and students alike, this book opens up important new perspectives on how we understand global creativity.

This book was published as a special issue of *Continuum: Journal of Media and Cultural Studies*.

Albert Moran is Professor in the School of Humanities at Griffith University where he researches cross-border trade in TV formats; screen geographies, and Australian screen history. His edited and authored publications include over 20 books. His most recent is *New Television: Globalization and the East Asian Imagination* (Hong Kong University Press 2007).

Michael Keane is a Centre Fellow at the Australian Research Council Centre of Excellence for Creative Industries and Innovation (CCI) at Queensland University of Technology, Brisbane. Michael is author of *Created in China: the Great New Leap Forward* (2007).

Cultural Adaptation

Edited by Albert Moran and Michael Keane

Routledge
Taylor & Francis Group

LONDON AND NEW YORK

First published 2010 by Routledge
4 Park Square, Milton Park, Abingdon, Oxon OX14 4RN
605 Third Avenue, New York, NY 10017

Routledge is an imprint of the Taylor & Francis Group, an informa business

© 2010 Taylor & Francis

Typeset in Times by Value Chain, India

British Library Cataloguing in Publication Data
A catalogue record for this book is available from the British Library

ISBN13: 978-0-415-54343-9 (hbk)

Contents

Part IV: Consuming Cultures

Part V: Framework

Notes on Contriburors

Jonathan Burston is Associate Professor and Rogers Chair in the Faculty of Information and Media Studies, University of Western Ontario, London, Ontario, Canada.

Bert de Muynck is an architect, writer and co-director of MovingCities. Since 2006, he has lived and worked in Beijing, China. His writings have been widely published. From May to July 2007 he was one of the coordinators of the Transdisciplinary Research on Creative Industries in Beijing Mobile Research Laboratory.

Anthony Y.H. Fung is an associate professor in the School of Journalism and Communication at the Chinese University of Hong Kong. He received his Ph.D. from the School of Journalism and Mass Communication at the University of Minnesota. His research interests included political economy of popular music and culture, gender and youth identity, cultural studies, and new media technologies. His new books are *New Television Globalization and East Asian Cultural Imaginations* (coauthored with Keane and Moran, 2007) and *Global Capital, Local Culture: Transnational Media Corporations in China* (2008).

Callum Gilmour worked as a Research Assistant at the Centre for Cultural Research, University of Western Sydney between 2007 – 9. He is currently completing his PhD in the Faculty of Creative Industries at the Queensland University of Technology. Callum's publications include articles in American Behavioral Scientist, Soccer & Society, and International Journal of Sports Communication.

Gerard Goggin is Professor of Digital Communication and deputy-director of the Journalism and Media Research Centre, University of New South Wales. His books including *'Internationalizing Internet Studies'* (2009; with Mark McLelland), *'Mobile Technology: From Telecommunications to Media'* (2009; with Larissa Hjorth);, and *'Cell Phone Culture'* (2006).

Bill Grantham is an independent scholar and lawyer based in Los Angeles. He wrote *'Some Big Bourgeois Brothel': contexts for France's culture wars with Hollywood* (2000).

Michael Keane is an Associate Professor and Centre Fellow at the ARC Centre for Creative Innovation (CCI) at the Queensland University of Technology in Australia. His chapter draws on research for an ARC Discovery Project 'Governance, human capital and regional investment in China's new creative clusters'.

Micky Lee is an Assistant Professor in Media Studies at Suffolk University, Boston. She has previously taught at Ithaca College, New York and the University of Oregon where she received her PhD in 2004. Her research interests are international communication, and the political economy of telecommunications, new ICTs, and information. Her writings have appeared in *International Communication Gazette, Feminist Media Studies* and *International Journal of Media and Cultural Politics.*

Anthony May teaches at Griffith University, Nathan, Queensland. He is the co-author, with Cory Messenger, of *Pop Up. A History of Pop Music since 1945* (Intellect, forthcoming).

Kelly McWilliam is a Lecturer in Communication & Media Studies at the University of Southern Queensland. She is the co-author, with Jane Stadler, of *Screen Media: Analysing Film and Television* (Allen & Unwin, 2009) and the co-editor, with John Hartley, of *Story Circle: Digital Storytelling around the World* (Blackwell, 2009).

Toby Miller is Professor of Media & Cultural Studies at the University of California, Riverside. His teaching and research cover the media, sport, labor, gender, race, citizenship, politics, and cultural policy. Toby is the author and editor of over 20 volumes, and has published essays in well over 100 journals and books.

Albert Moran is a Professor in the School of Humanities, Griffith University, Brisbane, Australia. He is the author of numerous books, including *New flows in global TV* (forthcoming, 2009), *Copycat TV: Globalization, program formats and cultural identity* (1998), *Wheel of fortune: Australian TV game shows* (with Chris Keating, 2003), *Understanding the global TV format* (with Justin Malbon, 2006) and the *Historical dictionary of Australian radio and television* (with Chris Keating, 2007).

David Rowe is Professor in the Centre for Cultural Research (CCR), University of Western Sydney, and from 2006 – 9 was CCR's Director. He has published on media and popular culture in many journals, including Media, Culture & Society, Social Text, Journalism, Social Semiotics, and Television & New Media. His books include *Popular Cultures: Rock Music, Sport and the Politics of Pleasure* (London: Sage, 1995) and *Sport, Culture and the Media: The Unruly Trinity* (second edition, Maidenhead, UK: Open University Press, 2004).

John Sinclair is ARC Professorial Fellow in The Australian Centre at The University of Melbourne. His published work over the last two decades deals with various aspects of the globalization of the media and communication industries, with a special emphasis on advertising and television, and covering Latin America as well as Asia and Australia.

Jinna Tay is a Research Fellow with the Centre for Critical & Cultural Studies, at the University of Queensland. She is working on an Australian Research Council, Federation Fellow project with Professor Graeme Turner on a comparative Asian television region. Her doctoral research focuses on the study of fashion, journalism, fashion magazines, and cities and her research interest ranges from popular culture, cultural consumption to media and cultural studies. Her chapter emerges out of that study. Her most recent publication is, *Television Studies After TV,* co-edited with Prof Graeme Turner.

Dr Rowan Wilken is a researcher and lecturer in media and cultural studies at The University of Melbourne. In addition to his work on advertising, a further key strand of his present research is concerned with exploring the interconnections between ICT use and social and spatial theory, especially in relation to mobile phone use.

Richard Woolley is a Postdoctoral Research Fellow at the Centre for Industry and Innovation Studies (CINIS) at the University of Western Sydney (UWS). Richard has a PhD in sociology and has been working on the training, collaboration strategies and other career dynamics of Australia's scientific research elite. He has been part of a research team studying scientific mobility in the Asia-Pacific region, publishing and presenting results from this study internationally. Richard also has an interest in economic sociology, particularly in relation to the construction of commercial systems through the deployment of knowledge and distribution of technologies. In this area, he has researched and published on the formation and governing of consumption markets including the commercial gambling industry.

Abstracts

Global franchising, local customizing: The cultural economy of TV program formats

Albert Moran

Contemporary international television offers a rich site for the investigation of matters concerning cultural adaptation. Over the past 20 years, a formalized, organized system has developed whereby program production knowledge can be borrowed from place to place for the re-creation of a television program in another territory. The TV program format is a kind of template or recipe whereby particular industry knowledges are packaged to facilitate this process of remaking. This article provides a trade background to the development of the TV format industry. It links the TV format's emergence to the practice of franchising, with its attendant cultural need to customize the format to suit local audience taste and outlook in a particular territory. This process of localization is examined on three levels using a model derived from translation theory. The article finds that the localization which occurs in such processes primarily involves the development of content that is nationally unexceptional through which audiences in a national territory can be addressed as a collective 'we'. Even beyond this detail, format adaptation raises crucial issues concerning globalization and nationalization, and these are addressed in the final part of the analysis.

'Dancing with my darlin'': Patti Page and adaptation in pop music

Anthony May

This article looks at the circumstances of adaptation in the pop music of the late 1940s and the early 1950s. Contrary to popular accounts that would see adaptation as evidence of an exploitative relationship based around race, the argument shows how adaptation functioned to generate and maintain growth in a three-tiered industry that was looking for new business models to adapt to the post-war commercial and social environments. Without denying the exploitative nature of adaptation, the article recognizes its value and strategic utility in the process of growth during a boom period that allowed unprecedented growth across all sectors of the industry, both mainstream and marginal.

Romance in foreign accents: Harlequin-Mills & Boon in Australia

Kelly McWilliam

This article is broadly interested in the adaptation and circulation of the mass-market romance genre as one example of the publishing industry's production and distribution of cultural artefacts within and across national borders. To consider this, the article focuses on the most successful mass-market romance publisher in the world, Harlequin-Mills & Boon, to ask the following questions: how has Harlequin-Mills & Boon, but particularly its international expansion into and operation in 'foreign' markets, been key to the contemporary success of the genre? What are some of the key strategies of the publisher's adaptation of the genre to new national markets, particularly in terms of issues of generic repetition and difference? What can Harlequin-Mills & Boon's negotiation of one national market, namely the Australian market, reveal about these questions in more detail? And how has the Australian office's recent shift from importing international content to commissioning local content signalled a critical shift in its adaptation of the genre to the national market? Ultimately, this paper proposes that these changes signal the publisher's entrée into the creative economy and the Australian office's shift away from being a branch office and towards being a creative branch.

Strategic regionalization in marketing campaigns: Beyond the standardization/glocalization debate

John Sinclair and Rowan Wilken

While the economic logic of globalization might impel global marketers to seek the economies of scale and other theoretical advantages of standardization, experience with the realities of linguistic and other cultural differences has obliged them to go some distance towards the 'glocalization' of their marketing campaigns. By examining the marketing strategies of Coca-Cola, McDonald's and Procter & Gamble over the last decade or so, with particular attention to Asia as a region, this article suggests that strategic regionalization in its various forms represents a kind of practical compromise with an extreme nation-by-nation approach – that is, a means of ensuring that campaigns are not glocalized any more than is strictly necessary. More generally, it points up the degree to which the global – local dialectic is, in practice, mediated not only by the national but also by the regional.

Recombinant Broadway

Jonathan Burston

There have been important changes in the way adaptations have been produced on Broadway over the last 30 years. Earlier Broadway-to-Hollywood transfers like *West Side Story* were produced in the economic contexts of mid-twentieth-century Fordism. Today, the transfer often works the other way, with Broadway adaptations of Hollywood blockbusters like The *Lion King created* and maintained from within a set of post-Fordist logics that live or die by the buzzwords of synergy and convergence. Not surprisingly, media capital has a larger role to play here than previously. This piece examines the changing conditions of production on Broadway that are the result of media capital's arrival on to the field of live-theatrical production since 1980. These changes, which have occurred in other established and emerging theatrical economies, feature levels of standardization, routinization and workforce alienation that are new to the theatre. Against any impulse towards formalism, this paper suggests that Broadway purists are often unfairly intolerant of adaptations. Under the theatre's new conditions of global-industrialization, a hit show's generic origins – literature, the movies, television – do not matter nearly as much as the newly mediatized methods of its global reproduction.

Global sport: Where Wembley Way meets Bollywood Boulevard

David Rowe and Callum Gilmour

As sport has circulated around the globe, its practice and expression have both replicated its established (and substantially Western-dominated) power formation, and to varying degrees challenged and modified it. The growing popularity of these sportsentertainment cultures reflects emergent media and leisure economies combining global aspirational cosmopolitanism with local cultural identities and histories. These multiple modes of cultural adaptation are evident in various Asian contexts in relation to the English Premier League (EPL) (and association football (soccer) in general), and hybridized local forms of global sport such as the Indian Premier League (IPL) cricket and the J-League (Japanese professional soccer). This article examines, in reference to these cases, the ways in which global sport culture is constrained by its historical inheritance yet also making new, multi-faceted cultural histories. It argues that the trajectory of sport's global cultural development is not subject to a simple logic of adoption and imitation, but is created out of multiple, intersecting correspondences – and non-correspondences – between histories, sites and social formations. The cultural complexion of sport – not least in the vibrantly performative sphere of fandom – is undergoing profound, culturally adaptive change, and analytical frameworks foregrounding sporting cultural importation and pale imitation require a concomitant re-consideration and adaptation.

Commercialization and culture in Australian gambling

Richard Woolley

Gambling has a prominent place in Australian culture. Following the liberalization of commercial gambling in the 1970s, Australia entered into an intensive phase of industrializing gambling as a form of mass entertainment and a significant source of public revenue and private profit. An important part of the industrialization of gambling has been technological transformation, which has made possible the production and distribution of new and enhanced gambling commodities. Knowledgeable actors in betting markets have always been required to adapt themselves to contingencies in assessing risks. These contingencies now include those that arise from institutional arrangements designed to protect commercial house advantage or provide credible gambling products to a mass market. This paper analyses the example of controversy surrounding betting on AFL matches, to shed light on dynamic interactions between gambling consumption, the introduction of technology and gambling commodities, and the culture of the game.

Localizing a global amusement park: Hong Kong Disneyland

Anthony Fung and Micky Lee

We examine how the Disney Company localizes Disneyland in Hong Kong. Unlike most transnational corporations that adapt local culture to accommodate local tastes in foreign markets, we argue that Hong Kong Disneyland adapts to the Chinese market by retaining and bringing to the fore Disneyland's global form. Evidence gathered from multiple visits to the park shows that local residents and visitors from Asia understand how global consumption works. For tourists from mainland China, however, the park offers a chance for them to assume the role of aspiring global consumers, an experience that state control prevents them from having in their daily lives.

Architecture on the move: Urban and architectural design in Inner Mongolia

Bert de Muynck

In Ordos, located 80 minutes' flying time west of Beijing, a new city is being built from scratch. With an expected population of 150,000 inhabitants by 2020, this development will test the idea of what it means to be Chinese in the twenty-first century. Interestingly, the cultural industries have become a part of the urban program. Located in the northeast of the city, the Ordos Cultural Creative Industry Park will become a testing ground for different intertwining aspects of the cultural industries in China. A part of the development is already unique: throughout the first half of 2008, 100 international architects from 27 different countries were invited to Ordos in order for each to design a 1000 square metre villa in the desert. This article discusses the creation of the Ordos Cultural Creativite Industry Park from different angles, touching upon the urban discussion in China, the policy of cultural industries, the establishment of creative business districts and the experience of foreign architects operating in China.

Great adaptations: China's creative clusters and the new social contract

Michael Keane

The transformation of China's urban landscape has witnessed a boom in cultural adaptation, namely the adaptation of a Western idea, the creative cluster. This chapter examines the formatting of hundreds of creative clusters-art centres, animation bases, cultural zones, and incubators. The cluster has important implications for how we understand China going forward into the second decade of the 21st century. The cluster phenomenon has resulted in to a substantive remaking of the social contract, between officials, entrepreneurs, local residents, academics-and most significantly cultural producers. However, these processes of adaption are mostly driven by real estate developers working in partnership with local government officials. Cut and paste design is the fast road to completion. In this sense, the description 'creative' may well be redundant.

Adapting the mobile phone: The iPhone and its consumption

Gerard Goggin

In this paper, I look at the Apple iPhone as a fascinating instance of adaptation, especially as it relates to digital cultures. A theme in the rise of the mobile, or cell, phone has been how it underscores the active role that people play in the orchestration and use of culture. The gambit of the iPhone is that the mobile phone itself will be decisively remade, and through this that media culture will itself be reformed. To make sense of this rapturous reception, I examine the iPhone as a notable instance of consuming culture. The paper discusses the double sense in which the iPhone functions both as a signal adaptation of the mobile phone at the same time as it introduces new practices and politics of adaptation.

'Pigeon-eyed readers': The adaptation and formation of a global Asian fashion magazine

Jinna Tay

Fashion magazines are the perfect medium to observe the twin synergies of adaptation and copying within the larger thesis of 'glocalization'. For it would seem that fashion magazines all over the world resemble each other to a certain degree but each publication is differentiated enough to sell monthly. While magazines overall tend to conform to a format and genre, individual publications have to shape themselves into unique titles with their own ethos and style that target a specific readership tribe. How do fashion magazines, local and syndicated, establish their niche and attract the readership they want? This article looks at a unique global fashion magazine, *WestEast* magazine from Hong Kong, and the strategies it has utilized to sell to a global readership from Asia to the West. This analysis of *WestEast*'s inception and growth demonstrates that the process of adaptation and copying begins with product conceptualization, and that a culturally nuanced understanding of its local readership and market is essential for its success.

Craic in a box: Commodifying and exporting the Irish pub

Bill Grantham

The Irish pub has become a worldwide commercial phenomenon. Its huge success since the 1990s has been the result of a conceptual, marketing, design and branding effort to create a complex figment of 'Irishness' attractive to non-Irish consumers, but increasingly difficult to reconcile with contemporary, multicultural Irish life.

Introduction: The global flow of creative ideas

Albert Moran and Michael Keane

VINCENT: But you know the funniest thing about Europe is?

JULES: What?

VINCENT: Little differences. I mean they got the same shit over there we got here but there it's a little different.

JULES: Examples?

VINCENT: Alright. Well you can walk into a movie theatre in Amsterdam and buy a beer. An' I don't mean just like no paper cup. I'm talking about a glass of beer. And in Paris, you can buy a beer at McDonald's. An' you know what they call a Quarter Pounder with Cheese in Paris?

JULES: They don't call it a Quarter Pounder with Cheese?

VINCENT: Naw. They got the Metric System. They wouldn't know what the fuck a Quarter Pounder is.

JULES: What do they call it?

VINCENT: They call it Royale with Cheese.

JULES (Savouring the phrase): Royale with Cheese!

VINCENT: That's right.

JULES: What they call a Big Mac?

VINCENT: Big Mac's a Big Mac but they call it Le Big Mac.

JULES (Imitates French accent): Le Big Mac. (Laughs) What they call a Whopper?

VINCENT: I don't know. I didn't go into Burger King. You know what they put on French Fries in Holland 'stead of ketchup?

JULES: What?

VINCENT: They put mayonnaise.

JULES: (Laughs)

VINCENT: I seen 'em do it man. They fuckin' drown them in that shit. (Tarantino 1994)

In the era of late modernism, various pressures play a decisive role in shaping the texture and meaning of the world around us. Population, work, transportation, new technologies of information and communication, lifestyle cultures and other forces are increasingly mobile, and this in turn helps make for a new set of public and personal surroundings. Social life everywhere now appears to share more and more in an international (if not a global) order, even if inequality and stratification remain common inside territories and across territories. Still, the perception is that a particular cultural life is increasingly universal. More and more consumers come to share in its practices and products, with those products becoming more and more homogeneous. This standardization argument finds much support in the apparent internationalization of many elements of media, entertainment, leisure and lifestyle cultures, with cultural conglomerates determined to maximize their global market reach. Once upon a time, in order to understand the economic, political and cultural forces affecting citizens and society, it was mostly deemed sufficient to look within the boundaries of the nation-state. Over the past two decades, these same pressures of globalization have impacted on critical research, highlighting the methodological need to adopt an optic that is more cross-border and transcultural as a means of gaining greater understanding of cultural life.

Notions of cultural adaptation

Currents of cultural contact across time and space are not new in human society (Lotman 1990). Many examples of various historical influences on the development of innovative forms, styles and genres of social institutions, practices and technologies come to mind, frequently in terms of the way that such customization has been analysed and understood (Bakhtin 1981, 1986). There is an obvious corresponding necessity linked with this imperative, namely to disaggregate the myriad ingredients of what might be called the whole way of life of people in the present. Human experience in the everyday world is shaped by different commonalities of social life. These are industrially organized and commodified forms of practice and experience at one end of the spectrum, and individually bound forms at the other end. An emerging global culture (if that is what is taking place) seems far from being undifferentiated. Tomlinson (1991) has convincingly demolished arguments concerning a developing nightmare of cultural imperialism. The latter notion was the idea that the formative cultural order would be constituted by a way of life that was universal and homogeneous (Ritzer 1998). Instead, there is more saliency in adopting Michael Tracey's (1988) characterization of the cultural landscape of television as a 'patchwork quilt'. This has vital bearing on our insight into how worldwide culture develops and changes. It supports notions of structure but also those of agency. Storper (2002) has usefully referred to the idea of 'knowledge flow' wherein what is being transmitted across space in those processes whereby culture is constantly being renewed and reinvigorated by what is being transmitted across space. Such processes are not confined to data and information transfer. Instead, knowledge flow refers to those bundles of creative ideas, practices, modes of thinking, customs, agendas, routines and so on that facilitate and expedite effects, outcomes, results and problem solving in other places and at other times (Moran 2009). By the same token, scientists and engineers talk of technology transfer where what is exchanged is not confined to machinery and material objects but includes any know-how which, when adapted, helps overcome difficulties and solve problems (Moran 1998).

Cultural adaptation is a convenient way of alluding to this important capacity of groups and individuals to shape the pattern of life of themselves and those around them. We understand the phrase to refer to the reorganization and rearrangement of popular culture, entertainment, consumption, creative design and the like on a large, even global, scale to fit the needs of particular situations, peoples, places and times. Originality, creativity, invention, breakthrough and discovery are not in question. Rather, the key dynamic of this flow of cultural ideas from one place to another and from one time to another involves multifarious processes of identification, selection, adaptation, possible rearrangement and redeployment of cultural forms and styles, often in unexpected and highly productive circumstances (as well as in more conventional and predictable situations). Of course, larger economic and political forces often drive such agendas, although it is a mistake to equate such flows of ideas with such pressures.

Big themes and little themes

Although the contributors to this collection engage with the matter of cultural adaptation in various arenas and in different geographical settings, nevertheless several overriding matters can be identified here. Sinclair and Wilken refer to Ritzer's (1998) spectre of 'McDonaldization' as one approach to the marketing of consumption with its determinative results for the cultural life of populations. According to such an outlook, the larger imperatives behind this kind of cultural adaptation set in train standardization and homogenization of consumption. This is not an approach that is congenial to the writers who have contributed to this issue of *Continuum*. Rather, at least four different overriding characteristics can be traced in their articles. First, cultural adaptation itself is seen to be a complex multilayered process wherein particular effects and consequences are determined. Adaptation in one situation at one time may be quite different from processes of imitation at work elsewhere at a different point in time. There may be all the difference in the world between the complex of forces at work in the US pop music industry in 1950, resulting in more than 10 cover versions of a popular song, and the pressures at work in the present in the commodification and export of the idea and material trappings of the Irish pub.

A second kind of distinction that operates in the collection relates to matters of marketing on the one hand and reception and use on the other. Although this division is not always addressed explicitly, still the writers indicate constant awareness that the two are not necessarily coincident. Hence, the articles by Rowe and Gilmour and by Woolley specifically address the figure of the consumer in the shape of the Asian sports fan on the one hand and the gambler on the other. In fact, going one step further, we might add that consumers themselves do not constitute a single homogeneous group when it comes to the success or otherwise of a particular cultural adaptation. Thus Fung and Lee note that three quite different kinds of attendees pass through Hong Kong's Disneyland theme park, with different kinds of cultural expectations that pose perplexing dilemmas for the cultural adaptation or otherwise engendered in such an undertaking.

A third kind of distinction at work in the collection relates to the macro forces driving cultural adaptation. The cultural remaking under investigation is invariably part of a widespread commodity culture. For the most part, private commercial interests lie behind such undertakings. The names of several of the individual business enterprises are well known internationally, with brand names such as Coca-Cola, McDonald's, Manchester United and Harlequin-Mills & Boon figuring prominently, as do their larger business institutional contexts including Hollywood, the US music industry, English Premier League football and Broadway. However, the nation-state itself may constitute an important alternative actor in particular scenarios concerning cultural adaptation. Political governmental forces are often shadowy but real presences in such matters as copyright and trade in the most market-bound of circumstances of cultural adaptation, but may become major overt players when larger scale cultural developments are at stake.

At least one other major distinction should be mentioned here because it relates to matters of methodology employed by different authors in the collection. Broadly speaking, two different approaches to the investigation of cultural adaptation can be seen in these pages. These strategies are macrocultural and microcultural, respectively. The first yields grand, large-scale characterizations of particular adaptations in major territorial settings such as those of China on the one hand and Asia on the other. At this level, broad theoretical matters come into play such as questions of globalization, standardization and localization. However, this avenue is by no means the only way to approach the matter of cultural adaptation. Several other articles develop more intimate case-study approaches that concentrate on the various interconnected elements at work in particular adaptation situations. These include specific financial imperatives, micro histories, technology, divisions of creative and technological labour, ideological outlooks and cultural gatekeeping.

One reason for the twin approach just outlined has to do with the general lack of any enquiry that explicitly, let alone comprehensively, addresses the dynamics of modern cultural practices, providers, purchasers and populations investigated here by our contributors within our theme subject. The social fields affected by creative idea flows is far reaching, and this issue of *Continuum* is concerned to help map such business and human undertakings. Labels such as mass media, culture industries, mass amusements and creative industries do not adequately address the range of instances that might be researched in terms of cultural adaptation. The present collection should be regarded as a down-payment on such a grand undertaking. Nonetheless, to facilitate such analysis, four particular cultural areas are addressed in these pages, following Vogel (2004). These divisions pay particular attention to three overriding dynamics relating to communications, commerce and consumption.

Media perspectives on cultural adaptation

The analysis concentrates first on media-based cultures, a recognition of the pivotal role of mass media in shaping the common stock of images and sounds in the everyday world. Mass media – whether sound, image or print – have the capacity to mediatize so much of social life so that it is imperative to begin the investigation of cultural adaptation by attending to this area. Albert Moran's article addresses the recent development and formalization of program remaking in the international television system. The format is the name given to the package of commercial knowledges that a devisor/owner makes available under licence to other production and broadcast companies elsewhere. However, there is always a chance that content developed for one audience may not suit the preferences and tastes of viewers in other territories. Hence, most remakes produced under this arrangement are deliberately varied and customized. Moran therefore traces the different levels at which such adaptation takes place, emphasizing that such 'localization' is always a 'banal' one.

Anthony May is concerned with adaptation in another branch of media systems, namely the pop music industry. In a closely argued case study, he focuses on a particular hit record and singer in the United States in 1950–1951 where the song was a 'cover' version of a musical number popular in the hillbilly market. The emerging US music industry was deeply influenced by Hollywood as far as adaptive-business practices were concerned, even if it had to apply such insight to quite different circumstances. Record companies chose songs for contracted artists for industrial reasons relating to their commonality and their copyright. The fact that a song was already known only assisted in its marketing by a record company in relation to a particular singer. Copyright in music and lyrics, but not in arrangements, further enhanced this commercial predictability.

Sinclair and Wilken's study of contemporary international marketing practices adopted by three US conglomerates – Coca-Cola, McDonald's and Proctor & Gamble – in Asia introduces other considerations regarding adaptation in media cultures. Faced with the problem of how to maintain and increase sales in different national markets, the three corporations have attempted to develop different accommodations between mass and differentiated business strategies. Market adaptation is inevitably determined within patterns of repetition and variation, and various market solutions are identified. However, calling attention to the limited choices that are made available to suit particular tastes, the authors prefer the term 'glocalization' rather than 'localization' when market differentiation occurs.

Kelly McWilliam examines developments in the oldest of the mass media. She is particularly concerned with the popular-romance novel, and especially, with the global marketing and publishing program of Harlequin-Mills & Boon. The latter has its origins in the United Kingdom and Canada, so its 'first language' has been English. Although the company has long been in the business of translating and culturally adapting its novels for readers in markets such as Sweden, its Australian branch had previously operated only as a distributor of its UK and North American titles. Since 2006, it has set about commissioning romances written by Australians as a means of producing a locally differentiated version of a global generic product.

Entertainment culture adaptation

Mass commercial culture can be organized around forms and meanings that are live, performance based, participatory and communal, so a second area addressed in this collection relates to adaptation in leisure and entertainment cultures. In this arena, the public is more evidently and visibly present, so its role and behaviour in live amusement events constitute an important part of what is occurring.

Jonathan Burston is concerned with the present-day, large-scale theatrical musical as it is staged in such districts as Broadway, the West End and their burgeoning satellite economies elsewhere. Over the past 30 years, giant media conglomerates such as Disney and Warner have moved well beyond their earlier Hollywood feature film platform to become important presences in other spheres of leisure consumption, including that of the live large-scale theatrical musical. Although Broadway and Hollywood have been in the crossover/adaptation business for a long time, there have been important quantitative and qualitative shifts in the way that these adaptations work. Now, Broadway adaptations of Hollywood blockbusters like *The Lion King* are created and maintained from within a set of post-Fordist logics offering new instances of standardization, routinization and workforce alienation.

Spectacle and drama are not unique to the stage. Recent years have seen various forms of sport become important components of a global entertainment machine. Rowe and Gilmour focus on two forms that have become especially important in Asia, particularly in India, namely English Premier League football and the Indian Premier League cricket. Their article is concerned with the spectacular, even exotic, fandom engendered by these two forms in that part of the world. Such fandom constitutes an intriguing cultural crossroads that is modernist as well as postmodernist, media bound as well as live and performative, native as well as cosmopolitan, authentic as well as hybrid.

Richard Woolley investigates the gambling leisure area, which is often closely tied to sporting events. Like Rowe and Gilmour, his stress falls on the figure of the consumer. He is concerned to emphasize the interaction of 'knowledgeable and interested gambling actors and the various forms adopted by the powerful commercial systems that manufacture and sell gambling commodities'. Wagering is constantly involved in various processes of cultural adaptation on both sides of this divide. Woolley develops an extended case study concerning the involvement of the Australian Football League (AFL) in a gambling franchise. However, he suggests that – especially late in the football season – internal pressures related to the football system may be at odds with competitive expectations regarding gambling on match outcomes.

Like Sinclair and Wilken, Fung and Lee in their discussion of Hong Kong's Disneyland also find it necessary to question the conceptual adequacy of the globalization/localization duality. Hong Kong Disneyland represents a cultural crossroads between different sectoral interests. On the one hand there is the Disney corporate interest with its determination to preserve the universal integrity of its various brands and franchises in the shape of characters, stories, imagery, music and sounds. However, visitors to the theme park are by no means socially and politically homogeneous. The experiences available to these consumers come to mean quite different things. In the face of the complex cultural phenomenon that is Hong Kong Disneyland, Fung and Lee suggest the need to reformulate present-day theorizing regarding cultural commodity adaptation.

Modifying public culture

A third arena of cultural adaptation relates to forms, arrangements, assumptions and priorities that evince a different phenomenal existence. This can be understood as the sphere of public culture where the phrase refers to a series of institutions that are physically durable, permanent, situated, frequently supported by government with state revenues and, notionally at least, accessible and available to the population at large. The first article in this part deals with architectural design on the move. Bert de Muynck is a writer who has followed the decision to invite 100 international architects to each design a villa at Ordos in Inner Mongolia. This desert region is very rich in energy resources so that the local council has embarked on an ambitious land development plan that would see the villas sold to coal and gas billionaires in this brand-new city appearing in the desert. ORDOS100 furnishes the chance for creative diversity where international ideas on housing can be tested under Chinese conditions of speed, quality of construction, labour skills, a creative industries context and return on investment. The result, de Munyck writes, may be a World Expo. Or Beverly Hills. Or an architectural zoo!

Meanwhile, Michael Keane examines the very recent take-up of the notion of cultural clusters in different regions of China. The creative industries cluster idea is a notion stemming from the United Kingdom. It envisages that the deliberate grouping of enterprises or industries has to do with media, artistic design, creative services and the building of copyright holdings generally, which are seen to potentially furnish an important growth engine for economic and cultural development generally. Creative industry clusters have sprung up across China over the past five years, often in abandoned factories, and represent the coming together of interests associated with local government, business entrepreneurs, artists, skilled and unskilled workers, tourist operators and others. Keane notes that, while creative industry developments may take several different forms, they represent a pragmatic solution to the problem of generating employment and growth in local economies. More importantly, he notes the degree to which the cluster idea is a deliberately formatted one so that its take-up has been considerably facilitated, hastened and guaranteed.

Lifestyle culture adaptation

Implicit across these milieus is the culture of human use and gratification. The last group of articles, 'Consuming Cultures', is especially interested in lifestyle cultures symbolized in retail goods and services. On the one hand these are seen to fulfil specific human needs – wants associated with food, drink, clothing and communications – and on the other to facilitate identity construction through celebrity promotion, product brand and lifestyle choice. Gerard Goggin's article on the iPhone analyses an emerging personal technology brought about by the convergence of mobile, computing, media and Internet cultures. Various forces of cultural adaptation are at work in this device, co-creating the iPhone, including its articulation within existing digital usages, its mediation by other cultural gatekeepers including Apple and various software developers and users, and its possible redefinition by hackers. Still, no matter how fluid the present moment of the iPhone is, the fact is that the technology is rapidly becoming intrinsic to everyday life as an aid to steering a path through personal choices and situations and as a means of signifying self and identity.

The second article investigates the manner in which the fashion magazine mediates lifestyle and identity cultures. Jinna Tay takes as case study the Hong Kong fashion magazine *WestEast*, which deliberately constitutes itself as a hybrid vehicle of Western (European) and Eastern (Asian) cosmopolitan consumption. The fashion magazine is a recognizable genre of publication with its emphasis on fashion photography, beauty products, feature articles on celebrities and lifestyle. *WestEast* represents a specific adaptation of this form that is attuned to the emerging cosmopolitan lifestyles of middle-class consumers in Asia. Although Hong Kong is perceived as the 'semiotic port' of Asian culture, nevertheless it was deemed necessary to first secure the magazine's appeal through its Western cultural framing before hybridizing and glocalizing it for Eastern readers.

Cultural adaptation is a restless creature, forever on the move. The last article in the section deals with the Irish pub commodity. Bill Grantham is careful to distinguish this international consumer product from its alleged parent in the shape of the public house or drinking tavern in Ireland. The latter is time bound, long in decline, and often now staffed and owned by Asians, Africans and Central Europeans. The Irish pub, on the other hand, is a boxed commodity, capable of limited customization to suit particular clients anywhere in the world, and promising to deliver a commodified Irishness in terms of setting, song and stout.

Toby Miller offers a highly perceptive and summative Afterword.

Acknowledgements

We are most grateful to the efforts of various people who have helped this volume come to successful completion. Our first debt of gratitude is to the contributors who have willingly responded to the theme, met our deadlines and additional requests, and produced a volume that is highly stimulating, learned and insightful. Sue Jarvis in Brisbane has been terrific in pulling the volume together for publication. Thanks to the team at *Continuum* and in particular Mark Gibson. We are also grateful to reviewers who generously provided critical feedback on the articles and to Joanna Kujawa for editorial support. Thanks also are due to Katherine Burton for supporting the publication of the collection under Taylor & Francis's special issue to book scheme, and to Stephen Thompson for providing editorial assistance with the production of the book.

References

Bakhtin, M. 1981. Discourse in the novel. In *The dialectical imagination: Four essays by Mikhail Bakhtin*, ed. Michael Holquist, 259–422. Austin: University of Texas Press.

———. 1986. The problem of speech genres. In *Speech genres and other late essays*, ed. C. Emerson and M. Holquist, 72–91. Austin: University of Texas Press.

Lotman, Y. 1990. *Universe of the mind: A semiotic theory of culture*. Bloomington: Indiana University Press.

Moran, A. 1998. *Copycat TV: Globalisation, television industries and cultural identity*. Luton: University of Luton Press.

———. 2009. *New flows in global TV*. Bristol: Intellect.

Ritzer, G. 1998. *The McDonaldization thesis: Explorations and extensions*. London: Sage.

Sinclair, J., E. Jacka, and S. Cunningham. 1996. *New patterns in global television: Peripheral vision*. New York: Oxford University Press.

Storper, J. 2002. *Globalization and knowledge flows: An industrial geographer's perspective in regions, geography and the knowledge based economy*. Oxford Subscriptions On Line.

Tarantino, Q., dir. 1994. *Pulp Fiction*. New York: Miramax.

Tomlinson, John. 1991. *Cultural imperialism: A critical introduction*. Baltimore, MD: Johns Hopkins University Press.

Tracey, M. 1988. Popular culture and the economics of global television. *Intermedia* 16(2): 19–25.

Vogel, Harold. 2004. *Entertainment Industry Economics: A Guide for Financial Analysis*. Cambridge, MA: Cambridge University Press.

Global franchising, local customizing: The cultural economy of TV program formats

Albert Moran

Introduction

Just 10 years ago, the *Big Brother* TV program format was first devised in the Netherlands by producer John de Mol for his production company Endemol. As Bazalgette (2005) has shown, the program was groundbreaking in terms of its premise of bringing together a group of young people and having them live in a confined space where they were filmed by a battery of hidden cameras. Starting in its home market of the Netherlands and shortly giving rise to a string of authorized spin-off productions across the world, *Big Brother* has become the most watched program in world television history. Its accumulated global audience for its different national versions regularly produced up to 2005 was estimated at 740 million viewers. It has also proved to be an extremely valuable franchise globally, generating more than US$10 billion in profits for its owner in the same period (Bazalgette 2005).

The phenomenal success of the program also heralded the maturation of an international system for the adaptation of television programs from place to place and from time to time. The borrowing and application of content ideas has been a recurrent characteristic of feature films, radio and television over many years. However, it has only been in the worldwide television industry that a formal, organized system of content adaptation seems to have emerged – even if these other audio-visual systems are following suit. It has done so under the rubric of the term 'TV format'. In this instance, the label refers to a method of practising television whereby a kind of unspecific, universal or de-nationalized program template or recipe is developed, which in turn can be customized and domesticated for reception and consumption by specific audiences in local or national contexts. The franchising of TV program formats is now a highly significant component of industry and cultural practice in modern television at both the national and international levels. Whether in the form of reality, game show, infotainment programming, makeover, talent show, sitcom or drama, the advent of the TV program format appears to signal the triumph of media globalization even while asserting the continued importance of local or domestic programming (Moran and Malbon 2006). What does this paradox imply? How is the TV program format best understood in the rapidly changing mediascape of present-day international television? What purchase does it offer us regarding the central theme of this issue?

This article explores the practice and meaning of TV format programming. Discussion is divided into six sections. The first part sketches the notion of a global television system. In the next two sections, I outline the main features of format programming, including its evolution as an industry practice. The following two sections develop an understanding of adaptation, drawing on semiotics and translation theory. In the concluding section, I consider this customizing of formats for home audiences under the label of nationalization.

Global television?

Since the late 1980s, television in many parts of the world has found itself in a new era (cf. Moran 2005). This is brought about by a unique intersection of new technologies of transmission and reception, innovative forms of financing, fresh means of imaging the audience, novel forms of content provision and new constructions of the television commodity. Technological digital convergence has become a dominant notion, with new players entering the distribution arena, including companies based in the telecommunications and computer sectors. Alongside the move to privatize much of television broadcasting services, there has been a parallel shift towards the use of independent production companies. Hollywood is no longer the dominant centre for the global production of popular television output. Other media capitals, including London, Hong Kong, Tokyo, Beijing and Mexico City, are becoming prominent hubs of a more international structure. Genres which once were marginal in popularity, such as game shows, talent contests, self- and home-improvement programs, and hidden camera 'documentary', have now become mainstream forms of programming output. Several of these program types highlight the capacity of the 'ordinary' person to go on television and are part of a neo-liberal zeitgeist. The abundant multi-channel services offered by technology share the same outlook. These are complementary to the information and entertainment provisions of broadcasters, and are increasingly more interactive than the older services. Producers and broadcasters are also finding fresh ways of tapping income.

The new structures of finance include not only the incidental commercial services available through domestic technologies of telephone and computer, but especially have to do with intellectual property (IP) rights.

This wave of development has been occurring with television in the West, but has also been apparent elsewhere. Many scholars see this increasing uniformity as evidence of an evolving global order in television (Waisbord 2004; Chalaby 2005; Tunstall 2008). In any case, though, it is worth recalling that an older term, 'world television', remains a valid label for such a configuration (Tunstall 1973; Parks and Kumar 2003, 14). A nascent world or global system of television has existed for some time, but may now be gathering strength. Chalaby identifies four interconnected components of a planetary television system that is complexly interrelated and changing (Chalaby 2005, 9). First, there is the media industry component, itself constituted by seven giant transnational television corporations. In addition, there are also other multinational organizations with strong regional sales or those with global reach, specializing in niche markets. Second, there is a technological infrastructure in the shape of worldwide communication networks that include cable, satellite and the Internet. This technical matrix facilitates and maintains the operation of these organizations. News and entertainment content, and associated data and services that circulate through the system, make up a third realm of global television. Finally, there is the world regulatory regime that includes various international bodies, technical and trade agreements, and legal decisions and ordinances. Such a scheme of a global television system is rudimentary, and various other interconnected sectors might and should be added. One such domain, for example, is the vital matter of television and media labour with associated issues such as employment, worker association, new technology, skill and craft matters which are in constant flux and contestation (Miller et al. 2001; Wasko and Erickson 2008). It should also be added that this system of a global television is, at best, nascent and emerging. Its reach is far from planetary, with many television viewers and television systems outside its operation (Tunstall 2008, vii–xii). Chalaby's own particular focus of interest is transnational satellite television. The latter is an important component of this complex evolving jigsaw of world television, but so too are other developments. One of these is especially important for this paper. It can be approached in terms of the content and related services that Chalaby identifies as the third of his four interrelated components.

Format programs

Aside from the home-grown provision of television output, two different forms of programming – distribution and production – now characterize international and national television (Waisbord 2004). The first of these modes is the more familiar and traditional. It has been labelled 'canned programming'. This form of content has to do with a program devised, produced and broadcast in one territory, which is shipped in cans or other containers for rebroadcast elsewhere. The canned program is always already nationalized, whether it has originated in the United States, the United Kingdom, Japan or elsewhere. When it plays in foreign-language territories, the canned program can be customized for home audiences up to a point by dubbing or subtitling.

The second, more recent mode is format programming (Moran 1998; Moran and Malbon 2006). Under this system, a program is devised, produced and broadcast in one territory. Subsequently, the program's format is made available as a set of services or franchised knowledges, which allow the program to be adapted and produced for broadcast in another territory. Various advantages attach to this kind of program remaking. Only successful, popular programs become the object of adaptation as foreign producers attempt to insure against ratings failure. The adaptation involves local production labour and can, for regulatory and other political purposes, count as local content. Mostly, the format program constitutes a flexible template or empty mould awaiting particular social inflexion and accent in other television territories to appeal to home audiences in that place. In this sense, the television format program might be said to be unbounded and universal. It offers itself as possible evidence of an emerging global television system.

As part of television format remaking, the program is usually modified in such a way as to seem local or national in origin. Whether or not the program is 'live' or scripted, the stories it tells will tend to deal with the audience's world. Moreover, its performers and participants will, for the most part, be ethnically familiar, speak one or other of the dominant territorial languages, be pictured in recurring, everyday locations dealing with recognizable situations, and behave in customary and familiar ways. In short, home audiences are seen to prefer a program that is attuned to their sense of who they are (Tunstall 2008). One might treat this form of program adaptation as one of format customization or indigenization. However, further qualification is in order as a means of sharpening our sense of the intriguing, paradoxical significance of program formats in the era of new television. For, on the one hand, they seem to have a go-anywhere quality that is not a part of the make-up of canned programming, while on the other they also appear to have a capacity to take root and nativize themselves in different television territories.

Format origins and development

Although TV format franchising reached its commercial zenith with *Big Brother*, nonetheless this form of program provision is not new. Elsewhere, for example, I have noted that the practice of translating the Bible from Latin into indigenous languages appeared in north-west Europe as early as the fifteenth and early sixteenth centuries (Moran 1998, 22). Similarly, the borrowing of radio program ideas for adaptation in other places occurred from the 1930s onwards. However, systematically facilitating broadcast program adaptation was something else, a new business departure. Take the example of the children's television program *Romper Room* (Hyatt 1997, 364). This first appeared on a local US television station in Baltimore in 1953. The year was an auspicious one as a US small business organization was beginning to undergo a transformation with the emergence of a new kind of franchising operation (Dicke 1982, 218–44). Hitherto, franchising had been organized around the distribution of material goods or products. In the 1950s a new kind of franchising emerged, one concerned with the distribution of service or tertiary products. One of the most spectacular examples of this kind of distribution arrangement was new fast food restaurant franchising, including Burger King, Kentucky Fried Chicken and McDonald's (Dicke 1982).

Although the Baltimore version of the *Romper Room* program was reasonably successful, its creators turned down an offer from the CBS network to buy the program. Instead, they packaged a range of relevant services and resources that they licensed out to a string of local television stations across the country. By 1957, 22 stations were licensing the format and producing their own local versions of the program. Six years later, 119 US stations had their own *Romper Room*. By then, the franchise was being distributed internationally and included several versions of the program being made in Australia and Japan.

Over the next half-century, the practice of TV program format franchising slowly came to be accepted in the international television industry. Since around 2000, format programming has been a mainstay of television output in many different territories across the world. It tends to operate in the higher budget end of markets and complements the import of canned programming. More importantly, the TV program format helps give popular transnational shows an indigenous or domestic look and sound. But how, exactly, does this format customizing come about? The next two sections examine this kind of adaptation, first in terms of two polar types of customization or translation and then in terms of the process of modification and variation.

'Closed' and 'open' adaptations

Adapting and producing a format program is usually an interactive process involving a centrally located format licensor and a licensee company based in a peripheral television territory. The former has extensive knowledge of the format and its inception in other places; they understand the pitfalls and difficulties as well as the potential triumphs and successes. The latter has a more intimate sense of the home audience culture, a greater intuitive sense of what will be suitable for viewers. A particular adaptation will mostly represent a compromise between these two parties. However, in rare instances, the licensing company will insist that the adaptation follows a standardized pattern that makes little or no concession to the interests and sensibility of a home audience.

With the international licensed adaptation and production of different national versions of *Who Wants to Be a Millionaire?*, for instance, the UK-based Celador, which owned the format, insisted that these adhere closely to the template created around the first incarnation of the program in the United Kingdom (Spenser 2006). This was certainly the case in India. To maintain the success of the program and to enhance the value of the brand and the franchise, Celador put a great deal of work into the collaborative activity of training local production personnel in the style and form it wanted in *Kaun Banega Crorepat*, the Indian version of the format. To this end, three staff members were sent out from the London office to India to train the local production team while four Indians went to the United Kingdom for further training (Spenser 2006).

This kind of TV program format adaptation might be called McDonaldization. However, it has generally been the exception rather than the rule. The experience of *Who Wants to Be a Millionaire?* has been repeated only with the international adaptation of one other TV format, *The Weakest Link*. In both cases, the licensed owners of the formats, Celador and the BBC, insisted that adaptations follow a highly standardized formula as far as the form and style of each adaptation was concerned (Jarvis 2006). This kind of decision making, where local input and inflection are discounted entirely, is unique in the international TV format business, although cases have occurred elsewhere in the culture industries. Burston (2000), for example, has reported a parallel process of international reproduction of cultural lookalikes in the case of large-scale live musical theatre in such markets as New York, London, Toronto and Sydney.

Elsewhere, drawing on semiotics and literary translation theory, I have labelled this kind of format program remaking situation a 'closed' adaptation (Lotman 1990; Lefevere 1993). What is required in any particular remake in any territory is a close approximation of the original version of the program. Head office has taken this crucial decision and local production personnel must follow this course of action. As in literary translation, the aim is to produce a 'literal' approximation of the original version of the format program. This lookalike, equivalent translation process emphasizes a high degree of fidelity to the original, even if the new version makes little concession to the interests and taste of a new audience.

Usually with TV program reproduction, however, a more 'open' adaptation is tolerated – even welcomed. Semiotics speaks of a more 'poetic' translation (Biguenet and Schulte 1990; Lotman 1990; Lefevere 1993). Further necessary changes have been introduced in a new version of a text in the interests of cultural audience intelligibility and access. In the case of TV program format adaptation, the creative sovereignty of the local production team is greatly enhanced in the process of adaptation and production. New versions of the format are likely not to be copies or lookalikes of earlier versions, and are far less substitutable one for another. Hence, for example, a loose adaptation of *Survivor* in China, *Into Shangrila*, seemed only a distant cousin to the format original, the Swedish *Castaway Robinson*, and to its most successful remake, the US program *Survivor*, although the latter two had more of a sibling resemblance to one another (Keane 2004).

Levels of format adaptation

As well as understanding this kind of transformation of formats in terms of a tug-of-war between serving a source on the one hand and serving an audience on the other, one can also grasp the process in terms of a trail of individual adaptation decisions about various elements of the format. When a TV format program is being tailored to suit a home audience, various levels of production determination come into play. Heylen (1994) has usefully suggested a tripartite scheme for understanding levels of activity in relation to a literary or written work that must be taken into account in translation. These involve linguistic codes, intertextual codes and cultural codes (Heylen 1994, 1–15). It is also relevant here to mention another tripartite model adopted in the analysis of another kind of cultural adaptation. This concerned the investigation of a Vietnamese newspaper in English conducted and reported upon by van Leeuwen (2006). The latter's schema again involved three steps. The first examined language decisions, the second focused on decisions affecting journalistic style and a third considered decisions affecting the cultural and ideological references in the source text.

Of course, television programs do not operate with linguistic codes. Instead, the poetics of television are located in matters of form and style (Bordwell and Thompson 2004). Style consists of staging, shooting, editing and sound. Form involves elements of extended organization and sequencing such as storytelling and magazine-type arrangements. These categories are complex and involve numerous individual elements that can be manipulated. This operation will be one of omission, inclusion, substitution or permutation. Elsewhere (Moran and Keating 2003), I have quoted from consultant Stephen Freeman regarding Grundy's changing the figure of the host for the Australian version of the game show *Wheel of Fortune*. This action is one of substitution wherein one element of the format is replaced by an equivalent element.

One other example of this kind of textual coding can be cited. Colour is a component of television's *mise-en-scène* that is frequently deemed to be nationally sensitive in game shows and other format genres. This element often necessitates cultural decision making in order to give a format program a recognizable 'look' as far as domestic audiences are concerned (Cousins 2006). Within the international television industry, however, there is a recognition that this authorization to vary cannot become a blank cheque. As another format consultant producer pointed out: 'Too much cultural tinkering ends up with a very bland show which, because of that will fail' (Spenser 2006).

Meanwhile, the orchestration of intertextual elements belongs to a second level of adaptation. These codes are not necessarily as discrete or self-contained as those just discussed. Instead, they appear to connect with specific bodies of knowledge held by particular communities, including both local production teams and segments of the home audience. At least three different sets of intertextual elements can come into play in a format adaptation. One set has to do with television production industries. Formats carry the imprint of institutions wherein they were first devised and developed.

Frequently, the adaptation and production of a format will necessitate a significant readjustment in a national television production milieu. Organizational norms, routines and practices that have perhaps been traditional to a local television production industry may have to readjust to a different regime that is more international, 'modern' and mainstream. A consultant producer may be pivotal in effecting such a change. This point about having to readjust local production conditions to help facilitate the remaking of a format adaptation was raised repeatedly in interviews. It underlines the point that the licensing of a format from elsewhere may trigger a cross-cultural exchange that begins with the readjustment of ways of working in television, whether by camera operator, editor, writer or others.

Intertextual knowledges can come into play in a second way in relation to the particular format program modes or genres that are being adapted. Elsewhere, I have suggested that the adaptation of a drama series format offers more opportunities for creative improvisation than does a game show format (Moran 1998, 107). In a similar vein, consultant producer Graham Spenser believes that game shows and single-issue talk programs are more international than other genres. With talk shows, though, it is a matter of 'how far you let the audience go with the subject' (Spenser 2006). Even with a game show, a good deal of preparation is essential to ensure a necessary accommodation between format and local culture. Various cultural screenings, such as those to do with particular national historical facts, come into play.

Third, a local adaptation may herald the introduction of a new television genre to a particular national television culture and industry. This has happened with the establishment of Australian soap formats in Western Europe in the 1990s, the inauguration of reality formats in many parts of the world after 2000, and the more recent introduction of telenovela templates in territories as diverse as India and Israel. Intertextual codes also operate in relation to performer persona knowledge in local television cultures. Hence, in the Netherlands and Germany, particular casting decisions soon brought about narrative deviations in the remakes of the Australian soap operas already mentioned (Kolle 1995).

A third level affecting format adaptation is that combination of factors that make for communal and national difference. Broadly, these include social matters of language, ethnicity, history, religion, geography and culture. They can usefully be referred to as cultural codes. Irrespective of the genre involved, a format program will carry particular situations, figures, subject-matters and issues. The extent to which these will be recognizable and acceptable within particular cultural settings may vary considerably. Hence, Graham Spenser saw no reason why *Survivor* would not work in Asia and the Middle East, even though no broadcaster in these regions had acquired the licence. After all, as he pointed out, *Who Wants to Be a Millionaire?* did enormously well in India. Frequently, though, regional or local culture will deem specific subjects and situations to be taboo and will not entertain them on local television screens. One dramatic case in point was the Middle Eastern version of *Big Brother*, which was broadcast to several national territories in the region. In Lebanon, according to Spenser, the local French influence allowed for a more tolerant reception but in Saudi Arabia there was great anger and outcry (Spenser 2006).

Gender relations and public behaviours can be another area of cultural sensitivity as far as format adaptation is concerned. A case in point was the relative failure of the format for *The Weakest Link* when it was adapted in Asia. With its domineering female quiz compère, this format presented an insoluble dilemma for its adaptation for that region. On the one hand, both the format owner and its Asian licensees wanted to retain intact the central ingredient of the dominant woman compère. On the other hand, such a figure had no cultural resonance in that part of the world. In the end, the combination of the alien figure and the public shaming of contestants was sufficient to ensure that various Asian versions of the format program were cancelled after a season or two.

More usually, though, successful domestication does occur with formats. In attempting to explain the 1990s success of the Polish adaptation of the 1950s US television sitcom *The Honeymooners*, one UK consultant specialist emphasized both a timeless element that might appeal to the Polish female audience as well as a more specific social appeal to a post-communist Polish audience (Cousins 2006). Of course, while producers frequently speculate about whether or not a format has undergone sufficient adaptation of the right kind, it is the home audience that is best equipped to deliver a verdict about its domestication. The same consultant producer told a story wherein the customization of *The Honeymooners*' format seemed so complete as to persuade one of the members of the local audience that the Polish program was not an adaptation based on an overseas format import, but was rather a program completely indigenous and home-grown to Polish culture:

> I was talking to a continuity girl and I said, 'Have you watched *The Honeymooners*?' And she said, 'Of course, it's very funny, I watch it every week. Why are you interested?' So I said that the company licensed out the scripts and she said, 'No, no, this is a Polish show.' This is the greatest accolade when they believe one of our shows is 'local'. (Cousins 2006)

The three levels of adaptation outlined are not separate in practice. Rather, most decisions in the process of adapting and producing a format program in a home territory are multilayered and shot through with many of the social values, ideological outlooks, cultural preferences and dispositions of a home community. To pay attention to these values is to tailor or customize the format for home audiences, always recognizing that this kind of construction of a television (global) localism or nationalism is a negotiated one. Like all nationalisms, it is an imagined one. The identifications it offers represent a matrix of particular choices that may suit the inclination and preferences of some members of the home audiences, while perhaps challenging, offending, marginalizing or even seeming irrelevant to the taste and sensibility of others in that population (Thomas and Kumar 2004). This 'cultural flexibility' in turn raises the matter of the larger social implications of TV formats and their cultural adaptation in particular territories.

National persistence

Up to this point, I have used several terms to indicate the modifications introduced when a TV format program is inflected towards the taste and sensibility of a home audience in a particular territory. 'Adapt', 'tailor' and 'customize' are deliberately neutral terms as far as their larger cultural implications are concerned. Even the label 'indigenizing' is only the loosest of social approximations when it comes to the translation of a format into an acceptable domestic idiom. In an era of increasingly larger movements of populations across the globe, it is impossible to designate a kind of majority television viewership of native origin and ancestry. Home audiences are likely to be mixed, heterogeneous and diverse in their interests and tastes. Van Leeuwen (2006) also points out that translation and adaptation are not synonymous, and should be applied to separate phases of the process of cultural transformation, whether it be the rendering of Vietnamese news into written English or the remaking of a television program for a particular territory. The term 'localize' also makes little real sense as a means of understanding processes taking place when a TV format is customized. The ambition is to gather the largest mass audience possible. Hence, even where specific choices have to be made regarding language, accent, ethnicity, religion and so on that will discriminate against various groups in a viewing population, format programming implicitly suggests that its address and appeal extend beyond local communities and attempts to talk to a national audience (Moran 1998).

In fact, a related difficulty with the label 'local' should be noted in passing. This has to do with its use in conjunction with the term 'global' as a way of encapsulating the dual spatialities of the TV format. This appears to be universal or global in its marketing and circulation even while it seems to be local and distinctive in its production and reception. Yet formats are not global in their circulation, although the international television industry claims that they are. While TV formats frequently achieve impressive numbers of licensings into many territories across the world, there are inevitably numerous other markets where they are not licensed, adapted or even noticed (Tunstall 2008, xi–xiv). Elsewhere, for instance, I have noted that Africa, parts of the Middle East, most of the former Soviet territories, and various parts of South and Southeast Asia are all sparsely represented at the international TV format fairs (Moran 2008). TV formats are better thought of as transnational or cross-border rather than global in their commercial circulation. Similarly, as already pointed out, formats are also better thought of as national rather than local in their cultural adaptation and appeal.

In other words, the advent of TV formats as a central element in the new television landscape appears to signal not the disappearance of the national in favour of the global and the local but its emphatic endurance or even reappearance. However, this endurance or reappearance takes place in a particular context at a specific point in history. The TV format industry's maturation as a mainstay of an international system of cultural exchange seems to point not to a strengthening of the global and the local at the expense of the national but to a reconfiguration of the national that may be to the detriment of those other two levels. As a transnational business system, TV formats are intimately dependent on the national – even if that persistence is frequently ignored and mistaken in favour of its own seeming disappearance. This apparent withering is strikingly in line with the postmodernist claim about the increasing irrelevance and disappearance of the national sovereign state (Hirst and Thompson 1995; Weiss 1998). But, just as the latter is a myth that suits various power interests in such areas as politics and economics, so the dissolution of the national into the local and the global as far as TV format adaptation is concerned also seems premature and even illusory. However, it is also worth ending by cautioning against any easy dualism or binarism regarding the global and the national. What the case of TV program format adaptation finally forces one to recognize is the recurring need to talk of various forms of the national as well as various forms of the global.

To understand this apparent vanishing, it is helpful to turn to a recent work of social science. Billig (1995) has usefully coined the term 'banal nationalism' to indicate precisely this visible invisibility of an ideological superstructure in everyday life. The kinds of mundane, taken-for-granted representations of the nation that are found in particular incarnations of TV program formats are the means by which the nation is reproduced as a hegemonic form. This ideological project is constant, ongoing across a series of different fronts in many arenas of social life including that of television. At the same time, such work is subtle, unobtrusive, banal. Of course, most TV format programming appears to be about almost everything else other than the power within which the aura of nationhood exists. Nevertheless, nationhood continues to be inconspicuously suggested in the interstices of format adaptations – in a detail of colour, a quiz question, an outdoor setting, a story situation, an accent, a theme song, and so on. Billig sees these small unobtrusive gestures and details as so many daily unnoticed 'flaggings' or reminders of nationalism. As he puts it: 'Banally, they address "us" as a national first person plural; and they situate "us" in the homeland within a world of nations. Nationhood is the context which must be assumed to understand so many banal utterances' (Billig 1995, 175).

How better, then, to cloak this persistent nationalism in the case of TV format programming than by cultivating the myth of its dissolution into globalism and localism. Nationalism has constituted a bedrock of television in the past. As this exploration of the phenomenon and meaning of TV format commerce and culture has suggested, this task is by no means superseded by the cultivation of other formations. Instead, in an era of rapidly changing features of the television landscape, TV formats continue to anchor their adaptations in the ongoing reality of the national.

Acknowledgement

This is a revised and expanded version of the essay 'Reasserting the National? Program Formats, International Television, and Domestic Culture' that is forthcoming in G. Turner and J. Tay, eds., *Television Studies after TV* (London: Routledge).

References

Bazalgette, P. 2005. *Billion dollar game: How three men risked it all and changed the face of TV*. London: Time Warner.

Biguenet, J., and R. Schulte. 1990. Introduction. In *The craft of translation*, ed. John Biguemer and Raimer Shulte, vii–x. Chicago: University of Chicago Press.

Billig, M. 1995. *Banal nationalism*. London: Sage.

Bordwell, D., and K. Thompson. 2004. *Film art: An introduction*. New York: McGraw Hill.

Burston, J. 2000. Spectacle, synergy and megamusicals: The global-industrialisation of the live-entertainment economy. In *Media organisation and society*, ed. J. Curran, 44–63. London: Oxford University Press.

Chalaby, J.K., ed. 2005. *Transnational television worldwide: Towards a new media order*. London: I.B. Tauris.

Cousins, B. 2006. Interview with Albert Moran. FremantleMedia, London.

Dicke, T.S. 1982. *Franchising in America, 1840–1980: The development of a business method*. Chapel Hill: University of North Carolina Press.

Heylen, R. 1994. *Translation poetics and the stage: Six French Hamlets*. London: Routledge.

Hirst, P., and G. Thompson. 1995. *Globalisation in question: The international economy and the possibilities of governance*. Cambridge: Polity Press.

Hyatt, W. 1997. *Encyclopedia of daytime television*. New York: Billboard.

Jarvis, C. 2006. Interview with Albert Moran. BBC World, London.

Keane, M. 2004. A revolution in television and a great leap forward by China? China in the global television format business. In *Television across Asia: Television industries, programme formats and globalisation*, ed. Albert Moran and Michael Keane, 88–104. London: RoutledgeCurzon.

Kolle, R. 1995. Interview with Albert Moran. Grundy World Wide, Sydney.

Lefevere, A. 1993. Introduction. In *Translation/history/culture*, ed. Andre Lefevere, 1–13. London: Routledge.

Lotman, Y. 1990. *Universe of the mind: A semiotic theory of culture*. Bloomington: Indiana University Press.

Miller, T., N. Govil, J. McMurria, and R. Maxwell. 2001. *Global Hollywood*. London: British Film Institute.

Moran, A. 1998. *Copycat TV: globalization, program formats and cultural identity*. Luton: University of Luton Press.

———. 2005. Configurations of the new television landscape. In *A companion to television*, ed. Janet Wasko, 291–308. Malden, MA: Blackwell.

———. 2008. Cultural power in international TV format markets. In *Trans-border cultural production: Economic runaway or globalization?*, ed. Janet Wasko and Mary Erickson, 333–58. New York: Cambria.

Moran, A., and C. Keating. 2003. *Wheel of fortune: Australian TV game shows*. Brisbane: Key Centre for Cultural and Media Policy, Griffith University.

Moran, A., and J. Malbon. 2006. *Understanding the global TV format*. Bristol: Intellect.

Parks, L., and S. Kumar, eds. 2003. *Planet TV: A global television reader*. New York: New York University Press.

Spenser, G. 2006. Interview with Albert Moran. Imagination, London.

Thomas, A.O., and K.J. Kumar. 2004. Copied from without and cloned from within: India in the global television format business. In *Television across Asia: Television industries, programme formats and globalisation*, ed. Albert Moran and Michael Keane, 122–37. London: RoutledgeCurzon.

Tunstall, J., ed. 1973. *Media sociology*. London: Constable.

———. 2008. *The media were American: The US mass media in decline*. New York: Oxford University Press.

van Leeuwen, T. 2006. Translation, adaptation, globalization: The Vietnam news. *Journalism* 7, no. 2: 217–37.

Waisbord, S. 2004. McTV? Understanding the global popularity of television formats. *Television and New Media* 5, no. 4: 359–83.

Wasko, J. and M. Erickson, eds. 2008. *Trans-border cultural production: Economic runaway or globalization?* New York: Cambria.

Weiss, L. 1998. *The myth of the disappearing state*. Cambridge: Polity Press.

'Dancing with my darlin'': Patti Page and adaptation in pop music

Anthony May

Sometimes analogies retain a power to frame debates much longer than their initial utility. Pop music, often characterized as the younger, rebellious sibling of the adult classical canon, remains an example of this. The problem with this attitude is that it can restrict the possible descriptions of the music and its institutions to an eternal adolescence. In turn, the significance of pop music as a cultural form, while taken seriously overall, is likewise restricted to elaborating juvenile concerns – issues of identity, rebellion, tribalism – and the opportunity to understand it as a major cultural form and institution is lost.

This article will examine the importance of adaptation in the formation of what might be understood as a core style for a cultural form that seems to be more traditionally associated with ideas of novelty, capriciousness, exploitation and consumption-driven production. It will focus on the production of American pop music in its formative decade following the Second World War, the years in which the industry as it is known today came into being. To assist that focus, an examination of one of its most historically significant recordings, Patti Page's 'Tennessee Waltz', will be central to the discussion.

The adolescent analogy has not affected the proliferation of arguments about pop music and adaptation. Adaptation, in a particular reading, has established itself as a foundational myth of pop music. The myth, in its most reductive form, characterizes pop music as a bland and anodyne commercial music directed at rapid turnover and inconsequential entertainment until in 1955, a decade after the war, significant individuals, responding to the changing world in which they lived, fused hillbilly music and rhythm and blues through the agency of charismatic white performers. From this union, rock 'n' roll was born. From that point on, the relationship between rock 'n' roll (and its derivatives) and pop music became a central fixation for popular music criticism. While the source musics – hillbilly, and rhythm and blues – were seen to embody an authentic relationship with their source demographics – the country and African-American audiences – pop was characterized as a de facto inauthentic and primarily commercial music whose overriding concern was financial profit. The picture is complicated, and so the source of endless debate, as the new music develops and grows a generation of consumers that claim their own authentic relation to the hybrid that would be known as rock.

This story only bears repetition and examination in that, in its development, it partially outlines aspects of the series of issues that can help the understanding of adaptation in pop music, sometimes by focusing on them and sometimes by their exclusion. The two key figures in this foundation myth are Bill Haley and Elvis Presley – Haley for the mobilization of the new American youth and Presley for the stylistic appropriation of what had, until not long before his emergence, been called race music. The focus here will be on Presley, although the history of Haley's outfit will be called upon by way of highlighting omissions from the general story.

The story of Presley and his performance of rhythm and blues under the tutelage of Sam Phillips, the Memphis record producer, is the stuff of legend and is clearly expressed in the following famous quotation from Phillips:

> Man, if I can find a white person who can give the feel and the true essence of a blues-type song, black blues especially, then I've got a chance to broaden the base and get plays that otherwise we couldn't. (http://www.elvis.com.au/presley/articles_samphillips.shtml)

On the one hand it succinctly expresses the relationship of the new music to the old, but on the other it highlights the commercial role of the transformation. It is therefore useful in affirming two dominant aspects of adaptation in pop music, namely the cultural and the industrial.

In the case of Sam Phillips, Sun Records and Elvis Presley, they were working at a level of the industry that was highly restricted. Phillips had been running Sun Records and the Memphis Recording Service since 1950, and the scale of the operation was typical of an independent studio/label. Phillips had the capacity to record and release records but the access to national distribution meant that it was usual for him to sell his recordings to larger independents such as Chess in Chicago or Modern in Los Angeles which could release the black artists that he often recorded into sympathetic markets. Phillips had recorded one of the earliest rock 'n' roll records – sometimes cited as the first rock 'n' roll record, Jackie Brenston's 'Rocket 88' – in 1951. The record reached number one on the rhythm and blues chart on the Chess label. A white performer like Presley might have offered new possibilities, but he was still bound by the size of his operation. Any advance that he might have made at Sun was stylistic. Presley's commercial success came when Phillips sold his contract to RCA Victor at the end of 1955 (Guralnick 1994, 215).

It important to note that, although Presley's first RCA record 'Heartbreak Hotel' went to number one in January 1956, by November of the same year Presley had starred in his first movie, *Love Me Tender*, for Fox Pictures. The projected lifespan of the pop singer was somewhat limited in the 1950s and the extension of a career into the movies was a path that had been pioneered by singers since before the war. Haley, too, was assisted by the movies. Although his version of 'Shake, Rattle and Roll' sold a million copies in 1954, it was the 1955 MGM film *Blackboard Jungle* that reprised his previously unsuccessful 1954 recording of 'Rock Around the Clock'. Movies were more than a destination for particularly talented individuals. The connection with the recording industry was intimate.

Before that relationship between the film and music industries is reviewed, it will be useful to set down a sense of the historical significance of Patti Page's recording of 'Tennessee Waltz'. The song was released by Mercury in late 1950, and by the last week of December 1950 was at the top of the *Billboard* chart. It stayed at number one until the end of February 1951. The two months at number one does not indicate the significance of the song. Songs did stay at the top of the chart longer in those days, but Page's recording managed to be a multi-million seller as a record as well as being one of the last releases to prompt multi-million sheet music sales at the same time (Sanjek 1988, 316).

The recording is of critical interest because an examination of its release and the institutional environment into which it was released allows for the refinement of the understanding of adaptation from being a merely exploitative process by corrupt individuals to an apprehension of a set of industry practices that underwrote a large part of the market as a whole. More particularly, adaptation can be seen to have been not just specific to rock 'n' roll but an ongoing set of practices that preceded rock 'n' roll, 'Tennessee Waltz' and the Second World War, and that continues to the present day.

The song was originally written for the hillbilly market in 1947 by Redd Stewart and Pee Wee King, with the latter making his own recording. Roy Acuff recorded an early version of it as well. Although Patti Page's version on Mercury is the most famous recording of the song, it was by no means the only cover. It remains a standard of adaptation through to today. More significant than its longevity, for this argument, are the recordings that existed at the same time as the Patti Page version. Patti Page's version was undoubtedly the most successful cover of the song, but that did not stop other artists recording the song as well. Between November 1950 and January 1951, six artists had top 30 hits on pop charts with the song. The success of Page's recording of the song even led Pee Wee King to reprise his original 1947 version for a hit on the country and western chart in February 1951 (Ennis 1992, 203).

Of the five contemporary hit versions that all charted in the period during which Page was at number one, only one of the six majors was not represented. Guy Lombardo released a version for Decca, Les Paul and Mary Ford for Capitol, Jo Stafford put one out on Columbia and RCA Victor charted twice, first with The Fontane Sisters and then with Spike Jones. All this occurred between November 1950 and February 1951 (Ennis 1992, 204). What is significant here is that the ownership of the song, which resided with Acuff-Rose Publishing, was no barrier to five of the six major labels having a hit. For the competing majors, other versions functioned not so much as competition but as publicity for their particular release. This hints at the clear need to understand the development of consumer loyalty in the pop market.

Whether the attraction was Patti Page or the song itself is moot. Patti Page's role in the pop world, however, is also helpful in understanding the developments that made adaptations such a key part of pop strategies. Page was a 'girl singer'. Typically, 'girl singers' were a new type of solo artist that had come about with the demise of the big band. Page had sung at the front of the Jimmy Joy band until it disbanded. She had also fronted one of Benny Goodman's small groups. Fronting a big band was not necessarily a prestigious role. The fame resided with the bandleader and the vocalist was seen as just another band member. But the life of the big bands came to an abrupt end after the war. Phillip Ennis points out that: 'In December 1946, within a single month, eight of the nation's top bands broke up' (Ennis 1992, 131). The majors picked up a number of these band vocalists and began to carve out solo careers for them. Page was picked up by Mercury in Chicago.

Page's career at Mercury demonstrates salient points, not only with regard to the issue of adaptation but also in enlightening on how labels treated the relationship between singer and song. Between the end of 1948 and early 1951, Page released a dozen songs on the Mercury label. Of this twelve, four were original recordings of songs and eight were adaptations of songs from other than original sources. Of those eight, five had been released previously and three came from the movies or musicals.

The first of these, 'Confess' (1948), had already been released by Doris Day and Buddy Clark some months earlier. Day, a girl singer also making the transition from being a band singer, was a major competitor for Page. Page was a favoured singer at Mercury while Day was signed to Columbia. Individuals may have competed to further their careers but the labels also had a major stake in their singers gaining prominence.

The first song of 1949 was 'So in Love'. This was a Cole Porter song from the musical *Kiss Me Kate*. The musical was an adaptation of Shakespeare's *The Taming of the Shrew*, making Page's release an adaptation from an adaptation. The point here is that not that all Page's songs can be examined with regard to their provenance. The data regarding the provenance of songs used by the majors for their recording artists may be assembled, and a picture of usage based upon a variety of statistical samples may be obtained. From the small sample of Patti Page's career, adaptations clearly form a major part of her output. With regard to the concerns of this paper, however, two more pressing questions come to light with this knowledge. On the one hand, why was it possible to use the work of other, less mainstream artists in this way, and on the other, why would they want to do it in the first place?

The answer to the first question is straightforward. As Russell Sanjek (1988) notes: 'Any legal barrier to the practice was removed by the courts in 1951 ... in a verdict that declared musical arrangements were not copyrighted property and therefore not subject to the law's protection' (1988, 325). With no legal barrier to copying arrangements, rerecording was a completely unfettered activity. The extent of the copying, from rerecording with the same arrangement, and in some cases the same session musicians and engineers, and simply a new, more acceptable vocalist, to the writing of new arrangements to go with established melodies and lyrics posed no threat to profits other than the payment of songwriter and composer royalties. Tom Dowd, the famed engineer at Atlantic Records, put it this way:

> [T]he music we had been recording had become so popular, the major record companies would cover one of our songs, if it was a hit, with a white artist. LaVern Baker had been covered two times by Georgia Gibbs, and I was helping make the Georgia Gibbs records because Mercury would call me up ... and say 'Come over, we're making the same song with a different artist. Make it sound the same.' People were using our records for demos for white artists. (Moorman 2003)

If that accounts for the possibilities of adapting works, the question of why they would choose to adapt works is much more complex.

There are two ways to account for the question of why majors would choose to adapt works. The first would be to consider adaptation practices in related entertainment industries and how the music industry might have learned from adjacent industries; the second is to examine the practice as it developed within the music industry. The most obvious related industry to the popular music industry is the Hollywood film industry.

Even without considering the issue of adaptation, the popular music industry and the Hollywood film industry had been closely intertwined since the putting of sound on film. The same technological developments assisted both industries, and the personnel who developed and operated those technologies moved freely across industry barriers. The primary practice of developing properties also required an integration of the two industries. In particular, they both needed to maintain a healthy relationship with the American Society of Composers, Authors and Publishers (ASCAP). ASCAP was the performing rights society that was charged with setting royalty rates and collecting royalties on all public performances of copyrighted music. To play copyrighted music in a film or to have a record played on the radio, you had to deal with ASCAP.

Both Hollywood and the music industry were concerned with developing commodities for a mass audience and Hollywood had long been in the business of using adaptations as the basis for movies. The relationship between the industries should not be seen as imitative. The music industry had long adopted the practice of exploiting the relationship with Hollywood, both as a source for its popular songs and as a vehicle for its popular performers. There is no more successful an example of this than the career of Bing Crosby and his recording of 'White Christmas'. Crosby, a white jazz band singer who transformed his popularity on American radio into global film star status, first sang the song in *Holiday Inn* in 1942 (Rosen 2002, 32–3). It was the most successful record to be recorded until Elton John reprised his own 'Candle in the Wind', a 1973 hit, at the funeral of Princess Diana and re-released the single in 1997. The perennial nature of a Christmas song no doubt had a lot to do with Crosby's success, but in terms of this discussion, its genesis in the movies is highly relevant.

In writing the history of the Hollywood mode of production before 1930, Janet Staiger notes the increasing dependence in Hollywood on sourcing stories from already published works, be they plays, magazine stories or novels:

> In summary, about 1911, two practices were in operation which contributed to a shift to film narrative sources coming from published literature and drama. First, advertising, which knows the value of prior literary and dramatic successes, promoted the adaptation of famous works, sometimes played by the stars who had made them a success. Secondly, adjudication of copyright laws necessitated acquiring the rights to narratives to be filmed – and as long as those purchases had to be made, the companies might as well advertise any valuable ones. Negatively, manufacturers began declining freelance plots, and within a decade, freelancers either published their work in print first or worked through agents. (Bordwell, Staiger, and Thompson 1985, 132)

Staiger identifies two practices that developed out of Hollywood's use of adaptations, one with a focus on advertising and one concerned with copyright. The issue relating to advertising worked in two ways. On the one hand, feature films that were based on works which had already been successful in publication or performance, whether contemporary or classical, could rely on an existing audience for the narrative and factor that audience loyalty into the ambitions for the film. On the other hand, the success of a narrative's prior incarnation made a suitable core for an advertising campaign for a feature film. What should be recognized here, however, is that the historical moment that Staiger addresses is the one in which film is being transformed from a single reel story into the feature film, more akin to what we know today. The shift was one from a single-reel film that worked by its ubiquity and presence amongst a stream of other single-reel films that audiences would consume on a very regular basis. The shift to the feature film was the shift to a commodity that was presented as a unique item that could be promoted in its singularity.

The second issue, copyright, also had two aspects to its development. First, the experience of litigation had taught the film companies that scripts or stories with an untested provenance could prove extremely expensive. That experience led to the second point, which was that in establishing control of the resources necessary to the production of films which now required major capital investment, the cost of copyright certainty was minor compared with litigation after the release of the film. In this light, the negative aspect that she outlines – the declining of freelance plots – can be seen as merely a part of formalizing an industrial process.

There are similarities as well as differences when considering how adaptation was managed in the pop industry. The major similarity was the commonplace recognition of copyright issues in the question of adaptation. As indicated above, there existed copyright agencies to maintain a watchful eye over the illicit performance of music. The 1909 Copyright Act had recognized the right to license copyrighted music for the purpose of profitable performance. By 1911, the Société des Auteurs, Compositeurs et Editeurs de Musique (SACEM), the French performing rights agency, had opened for business in New York. By 1914, an American agency, ASCAP, had taken up the task of protecting royalties (Sanjek 1988, 37–8). The relationship between ASCAP and the movies was difficult at times, but this tended to be around the setting of rates rather than the principle of copyright protection. Both industries were, after all, well aware of the income generated by the protection of royalties. Through the intervening years, the pop industry had taken songs from the movies for its hits and the movies had used pop songs to play in their films.

The main issue when considering the different ways in which the two industries utilize adaptations is concerned with the media of transformation. Hollywood, for the most part, has a relatively straightforward regime of adaptation. It takes works from other media and adapts the story into a feature film. Those other media may be plays, novels or stories. In more recent times, they have begun to adapt from comic books and video games. Music maintains a different operation.

For the most part, the pop industry takes its source from a song in a different market and transforms it into a mainstream pop song. There are exceptions. Before she recorded 'Tennessee Waltz', Patti Page recorded 'All My Love (Bolero)', which was an adaptation of a French pop song. The French song had used 'Bolero', by Maurice Ravel, as its source and this was indicated in the title of Page's release. Apart from this difference – moving across media as opposed to moving within media – there is also a complication in that whereas Hollywood adapts a title from another form and promotes the film as a filmic version of that initial narrative, music works in a different fashion. In the case of 'Tennessee Waltz', while the song may have come from the hillbilly market and then, through the recordings of Page and others, achieved a recognition status of its own, in the first instance it was designed to promote the career of Patti Page as much as it was to highlight the specificity of the song. Songs are sold to the public as great songs but the primary concern of the major label is to promote the career of the singer. Adaptations are chosen on the basis of their compatibility with the singer rather than as something that might be promoted on their own behalf.

Beyond these similarities and differences, however, there is a more underlying reason for considering these two industries in light of their adaptation practices. In each case, the process of adapting material either from other media or other markets illustrates an objective that is common to both industries. As purveyors of mainstream entertainments, the properties that they transform are transformed into mainstream products. In the case of the Hollywood feature film, this is most clearly exemplified in the instance of translating a foreign-language film into a Hollywood feature. Apart from making accommodating shifts of location, scenario, idiomatic expression and other markers of US narratives, Hollywood features rely on a certain technical quality, editing styles, pacing, rhythms and other mechanical aspects that signal the sense of Hollywood feature to an audience. Likewise with pop music.

In taking a song from a regional market, a major label performs certain operations on the arrangement in order to place it with a mainstream singer. This can be considered in two ways: what is brought in and what is left out. Very often what is left out is that part of the original that is distinctive to the regional market but not essential to the song. In the case of 'Tennessee Waltz', there are some very clear shifts. The song is slightly slowed down. The introduction, a rendition of the melody on a country fiddle in the original, is replaced by a muted trumpet or trombone playing a brief fragment of the melody. The original features an accordion throughout that has a small solo spot. This is replaced by a sparse piano figure. The distinctive country sound of the lap steel guitar is taken away. The effects of these changes are to remove the country waltz feel from the song and to develop a ballad feel while retaining the waltz time. Perhaps the most significant shift is the double tracking of Patti Page's voice.

At the time of its release, the end of 1950, nothing would have signalled both the contemporary nature of the song and its technological advancement as the double tracking of the singer's voice. Patti Page was not alone in having her voice overdubbed to give the effect of a duet with herself, but it was a particular marker of her style at this time. Significantly, Les Paul and Mary Ford, who also had a hit with 'Tennessee Waltz' and a further hit, like Page, with 'Mocking Bird Hill', were also known for using tape technology to double track vocals. It was a novelty that signalled the experimentation that the mainstream was undergoing with tape recording equipment.

A good example of what could be left out can be seen in the career of Nat King Cole at Capitol. In this case, what should be considered is the adaptation of Nat King Cole the jazz singer, pianist and leader into Nat King Cole the balladeer. What is most striking, if his output is followed, is his role as pianist that falls away as his vocal style as a balladeer is emphasized. In this case, the performer has been adapted into a different and mainstream style.

The importance of the differences between the regional hit and the mainstream hit becomes more apparent when the issue is considered at an industry level rather than in the comparison of two individual recordings. During the period between the end of the Second World War and the beginning of rock 'n' roll in 1955, the industry had four operative levels. There were two levels of major record companies and two levels of independent companies. The simplest and most accurate way of discriminating between a major label and an independent label was the issue of distribution – the majors had their own national distribution arms and the independents did not.

The four levels of the pop music industry can be described in the following way: there were three pre-war majors that had their base in New York – RCA Victor, Columbia and Decca; three newer majors – Capitol, Mercury and MGM; a number of larger independents around the country that had access to nationwide distribution; and an even larger number of independents around the country that had access only to regional distribution.

The pre-war majors led the mainstream industry in sales and prestige. RCA Victor and Columbia were companies that had begun in the late nineteenth century with the development of wax cylinders for recording (Ennis 1992, 44). Decca was formed by the British Decca label in 1934. The first two were the only two majors to survive the introduction of radio in the 1920s and the three were the strength of the industry after it had come through the Depression and the war. The second level of majors comprised the newer companies formed in the 1940s. Capitol was the earliest, formed in Los Angeles in 1942. The company came into existence at the most inauspicious time, immediately prior to the 1942 Petrillo recording ban. With just enough time to stockpile recordings, Capitol survived the war and made it into the prosperous post-war period (Sanjek 1988, 217–18). MGM Records was formed in Los Angeles in 1946 to provide an outlet for MGM Pictures' soundtracks. Mercury was formed in 1945 in Chicago.

The two levels of independents are not so easy to identify by name. Looking at the rise of the independent label in the first decade after the war is to see a dizzying array of diverse labels of all sizes and with affiliations to all markets. Rick Kennedy and Randy McNutt

(1999) show the problem in a very simple statistic: 'From 1948 to 1954, about 1,000 new record labels were formed' (xvi). Independents had existed before the war and been responsible for some striking developments in popular music. Gennett Records, for example, had developed out of a piano manufacturer and formed as a label in 1915 (4). The label played a significant role in the recording of early jazz, particularly Louis Armstrong and Hoagy Carmichael. In the same way, many early independents have stories built around the expansion and contraction of related business concerns. These things changed after the war.

After the war, independents flourished. There are two reasons for this. On the one hand, the general rise in business opportunity born of a growing economy and an across-the-board rise in standards of living meant that a market with massively rising consumption rates could take a much increased level of business activity and competition. On the other hand, technological developments – in particular the introduction of tape recording – meant that smaller concerns could afford to maintain a recording studio and develop recording expertise. There are notable exceptions to this. While Chess Records of Chicago was one of the most significant post-war independents in the rhythm and blues field, in the years between the label's inception and the rock 'n' roll boom – 1948–1957 – Chess recorded most of its artists at Universal Studios in Chicago. The label did not have adequate facilities until 1958 (Cogan and Clark 2003, 120).

Although the general trend is clear – independents flourished in a growing economy – close attention needs to be paid to the manner of that blossoming. The romance of independent labels is archetypically American: the local entrepreneurial spirit rising up and making it in the big leagues of organized capital. The reality was somewhat different. In *The Death of Rhythm and Blues*, Nelson George (1989) prints a table of independent rhythm and blues labels that emerged between 1942 and 1952 (27–8). George accounts for 28 labels. The list is not exhaustive. George's point, however, is that in this distinctly African-American music:

> Most indies were started by whites, many of them Jewish; blacks weren't the only people kept out of the American business mainstream by discrimination. Unwelcome on Wall Street, many Jewish businessmen looked for places where there were fewer barriers to entrepreneurship. They often turned to black neighbourhoods – in some ways paralleling blacks' discovery that their avenues for advancement were less barricaded in the world of entertainment. (28)

The simple equation of music style with label owner is not possible, nor is the assumption that all new technology flows through the system at an even rate. There is a way of grouping independent labels that will help in understanding the processes of adaptation. George's list of rhythm and blues labels, while indicating a dearth of black owners, does show that for each of the labels that formed in New York or Los Angeles, one formed in some regional centre like Houston, Cincinnati, Nashville, Chicago, Memphis or Jackson. They spread across the country with their location determined by a range of factors, local musical activity being only one.

With a dispersed set of production centres, distribution became a key to success and is the yardstick to judge the type of independent label. It is worthwhile separating independent labels into those that had access to national distribution and those that were left to service only regional markets. National distribution was not something that an independent label could normally undertake alone. There were, of course, exceptions. Brian Ward points out that Syd Nathan was extremely proactive in building the independence of his Cincinnati-based King-Federal label. Nathan not only ran an independent national distribution arm of the company but also set up his own recording studio, publishing company, pressing plant for records and photographic laboratory to produce record covers and labels (Ward 1998, 25).

It was more normal for a record company to contract with a distribution company from amongst a number that had grown up to deal with this pressing issue for independents. Based in a variety of cities around the nation, independent distributors filled the gaps in the system (Ward 1998, 25). But not all independents could afford the access to markets that these services offered. Typically, distributors relayed records to the various enclaves or markets around the country that were sympathetic to the product. While the example here has tended to focus on the rhythm and blues market and its distribution through African-American communities across the country, the same logic applies to hillbilly music, klezmer, polka and a variety of minority and independent tastes.

The appropriate dividing line for independent labels, then, is between those that could afford a nationwide network of markets and distribution and those that were restricted to local distribution. The latter ranged down to label owners driving stock around in the back of the family car. This leaves a picture of a four-tier industry from the older majors (RCA Victor, Columbia and Decca), through the newer majors (Capitol, Mercury and MGM), the independents with national distribution, but perhaps limited other capacities, and the smaller independents that were restricted to regional markets. The movement of songs between the levels was almost entirely hierarchical, from the bottom to the top. It served little value for the independents to cover the songs of the majors, although if a regional artist could mine an audience loyalty that would make the enterprise profitable then economics ruled the day.

In light of the above, it is necessary to return to the question of the types of songs that were adapted. There was a qualitative difference between the recordings that circulated in the regional or independent markets and those that were released by the majors. Unlike the mythology and history of Elvis Presley, the phenomenon of one of the world's future industry-changing stars walking quietly into an independent studio and asking to be recorded was not common. More usually, independent studios and labels recorded local artists who were self-contained.

Independent recording artists were performing artists who had developed their material and their arrangements over a period of live performing. By the time they arrived at an independent studio they had a body of material that had been tested in front of live audiences and was very often dance based. This is not to say that they were dance tunes, but rather they were of a tempo and feel that dancers could enjoy. That was the performance environment in which they developed. This was part of their attraction. They performed music that at some level was exciting.

Major labels, however, were abstracted from this process. Major labels recorded recording artists. To return to the example of Patti Page, in shifting from a band singer on the road to a girl singer at Mercury, her career as a live performer was relegated to second place against her career as a recording artist. Whereas she might have developed her singing style in front of a big band, at Mercury she was produced by Mitch Miller, who had the resources of professional arrangers – he and Percy Faith – and professional session musicians. The ambition of the session was somewhat different. Unlike the independent session that was directed at making a hit single by capturing the excitement of the seasoned combo, the major label session was directed at making, with the experience of a seasoned singer, a hit record based on the style of other hit records.

There were, clearly, exceptions to this. Tom Dowd's testimony indicates times when the majors wanted to import the excitement of the new sound into the repertoire of the majors. The two strategies were consistent if viewed from an industrial perspective of maintaining the quality of product and developing the standard through importing new advances.

There was a third aspect of the core style of the majors, and that was the utilization of novelty. The technological development of tape recording was not used solely with regard to economy and fidelity. Many producers and engineers experimented with a range of effects that were tape based to distinguish their music. In the case of Patti Page and others, double tracking the vocal became a distinctive feature of her recordings. Neither was the practice of tape and other forms of experimentation restricted to the majors. Peter Doyle (2005) examines a range of techniques that circulated throughout all levels of industry recordings during this period, but notes that Miller's use with Page was restricted to double tracking the voice (144).

On the same recording, as noted above, the chorus-length country fiddle introduction to the song on the Pee Wee King original is replaced by what might be a muted trombone or trumpet. The indistinct quality of the instrument is not so much a deficiency as a deliberate obfuscation by Mitch Miller in the production of a novelty sound. A core style surrounded by novelty sounds and new sounds was the marker of the majors.

To return to the original myth that constructed adaptation in the post-war recording industry as a mode of rampant exploitation, it can be seen that a lot more was happening than just the theft of creative product. To see the picture more clearly, adaptation in pop music needs to be understood as a way of marking out pop music from the other popular musics of the day. The adapted songs were not so much reduced in their transformation as recrafted into a different kind of product, one that built the industry and formed the basis for the pop music that has continued through to today.

References

Bordwell, D., J. Staiger, and K. Thompson. 1985. *The classical Hollywood cinema: Film style and mode of production to 1960*. London: Routledge.

Cogan, J., and W. Clark. 2003. *Temples of sound: Inside the great recording studios*. San Francisco: Chronicle.

Doyle, P. 2005. *Echo and reverb: Fabricating space in popular music recording, 1900–1960*. Middletown, CN: Wesleyan University Press.

Ennis, P.H. 1992. *The seventh stream: The emergence of rock 'n' roll in American popular music*. Hanover: University of New England Press.

George, N. 1989. *The death of rhythm and blues*. New York: E.P. Dutton.

Guralnick, P. 1994. *The last train to Memphis: The rise of Elvis Presley*. Boston: Little, Brown.

Kennedy, R., and R. McNutt. 1999. *Little labels – big sound: Small record companies and the rise of American music*. Bloomington: Indiana University Press.

Moorman, M. 2003. Dowd, Tom. In *Tom Dowd & the language of music*. New York: Palm Pictures.

Rosen, J. 2002. *White Christmas: The story of an American song*. New York: Scribner.

Sanjek, R. 1988. *American popular music and its business: The first four hundred years*. Vol. III, *From 1900 to 1984*. New York: Oxford University Press.

Ward, B. 1998. *Just my soul responding: Rhythm and blues, black consciousness and race relations*. London: UCL Press.

Romance in foreign accents: Harlequin-Mills & Boon in Australia

Kelly McWilliam

Introduction

The publishing industry has been one of the principal institutions for the production and distribution of cultural artefacts, if not since the fifteenth century, when a nascent transnational publishing industry developed across Western Europe, then certainly since the sixteenth century, when the industry began to emerge as a genuine mass medium (Pettegree 2002, 80). But while the emergence of mass market publishing was certainly dependent upon the development of relevant printing technologies – which, according to Marshall McLuhan (1967), allowed the publishing industry to become the first industry in the world to mass produce cultural commodities (185–6) – the industry also relied, from the very beginning, on a suite of forms and practices designed to culturally court mass audiences. One of those forms and practices was genre.

Genre is a consensual system of categorization that privileges particular textual, intertextual and extratextual conventions – such as plot, setting, style, author, series, brand, and so on – over others; it is the 'primary logic for popular fiction's means of production, formal and industrial identification and critical evaluation' (Gelder 2004, 40). In a literary context, genre is discursively constituted in the tripartite negotiations between the publishing industry, its readers and the cultural mores of a given time and place. As it does in other media industries, it usually functions first and foremost as an industrial marketing device (McWilliam 2009a, 237–9). By emphasizing conventions over originality, genres provide both an *horizon d'attente* (horizon of expectation) for readers and a *modèle d'écriture* (model of writing) for authors within which 'broad patterns' are repeated across texts and nuances negotiated within texts in an ongoing constitution of genre categories (Holmes 2006, 6). Like television formats, then, literary genres 'constitute processes of systematization of difference within repetition' (Moran and Keane 2004, 200; Neale 1980, 51). But while genres are intuitively associated with the repetition of their conventions, the differences between genre texts are also 'essential to the economy of genre' (Neale 1980, 22–3). Where repetition produces familiarity and interest in a genre, differences between texts, or specific variations of genre conventions, create interest in individual texts, thereby perpetuating the genre. Nevertheless, some genres – but none more so than the mass-market romance genre – are thought to be more repetitive and 'convention-bound' than others (Gelder 2004, 43).

The romance genre

The romance genre is one of the most 'profitable of all popular genres' in the history of mass-market publishing (Creed 2003, 100). In fact, for Lynne Pearce (2004), it is the 'most popular of *all* stories' (536). The genre itself originally emerged in twelfth-century France – though some critics claim that prototypical models of the genre existed in ancient Greece – but only developed into its current mass-market form in the eighteenth and nineteenth centuries (Regis 2007, 4). Now, as then, the genre is defined by its thematic emphasis on love and its narrative focus on the re/formation of a (typically heterosexual) couple. The standard romance narrative, irrespective of sub-genre or cycle, progresses along three stages: the couple meets (or re-meets); they negotiate 'a series of obstacles both internal (psychological and emotional) and external (social and material)'; and they either unite (for a happy ending) or separate (for an unhappy ending) (Holmes 2006, 6; see also McWilliam 2009b). Pamela Regis (2007), however, claims that happy endings are not, as Diana Holmes suggests, one of two possible dénouements in the mass-market romance genre; instead, they are a necessity, because 'readers insist on it' (9).

Commercially, of course, the genre has experienced one happy ending after another. After being 'something of a phenomenon' in the Western publishing industry for the last three decades, or since the 1970s paperback revolution, it has reached still higher levels of success in recent years in the United States, United Kingdom, Europe and Australia (Eike 1986, 25; RWAust 2004). In 2006–2007, for example, romance was the largest selling fiction category in the United States, accounting for almost US$1.4 billion in estimated revenue; this was the equivalent of 40% of all popular fiction sales in the country and the majority – more than 50% – of all mass-market paperback sales (RWA 2007; McLean 2008). But what can this extraordinary market success reveal, if anything, about the dynamics of change at play in the romance genre within and across national borders?

Aims

This article is broadly interested in the adaptation and circulation of the mass-market romance genre. To consider this, I will discuss one publisher of the genre in one market. Specifically, this article focuses on the most successful mass-market romance publisher in the world, Harlequin-Mills & Boon, in order to ask the following questions: how has Harlequin-Mills & Boon, but particularly its international expansion into and operation in 'foreign' markets, been key to the contemporary success of the genre? What are some of the key strategies we can identify in the publisher's adaptation of the genre to new national markets, particularly in terms of issues of generic repetition and difference? What can Harlequin-Mills & Boon's negotiation of one national market, namely the Australian market, reveal about these questions in more detail? And how has the Australian office's recent shift from importing international content to producing local content signalled a critical shift in its adaptation of the genre to the national market?

Harlequin-Mills & Boon

As mentioned, in 2006–2007 romance fiction accounted for almost US$1.4 billion in revenue in the United States alone as the largest selling fiction category in the country; in those years, the top five publishers of romance fiction operating in the United States included Random House, the Penguin Group, HarperCollins, and Kensington (RWA 2007; McLean 2008). But the largest by far was Harlequin-Mills & Boon, earning over US$410 million in (romance) sales in the United States alone (McLean 2008). In fact, Harlequin-Mills & Boon is the largest publisher of romance fiction in the world, a distinction it has held since 1981 – or, extraordinarily, for more than two decades (McAleer 1999, 284–5). The company, with its stable of around 1300 authors worldwide, publishes more than 115 romance books each month in 26 languages and 109 international markets (H-MB 2006; see also Orr and Stout 2008). These international markets are managed through the principal offices the publisher maintains in Amsterdam, Athens, Budapest, Hamburg, Granges-Paccot, London, Madrid, Milan, New York, Paris, Rio de Janeiro, Stockholm, Sydney, Tokyo, Toronto and Warsaw, and the licensing agreements it holds with companies in at least nine other countries. The publisher also has 'outlets' around the world, including in 'Malaysia, Singapore, Korea, the Philippines, Thailand and Eastern Europe' (Creed 2003, 100). With this broad commercial foundation in place, Harlequin-Mills & Boon continues to grow: in 2005, the company sold 131 million books; in 2007, it sold more than 160 million with an estimated readership of 50 million, which is the equivalent of selling 5.5 books per second (H-MBA 2006; H-MBA 2007a). But how did this publisher come to so thoroughly dominate mass-market romance publishing, and what can its dominance reveal about its adaptive diffusion of the genre around the world?

Harlequin-Mills & Boon, as the dash suggests, is the result of a merger between two companies. Mills & Boon was founded in London in 1908 as an educational and general publisher. By the end of the First World War the company had established two strategies that would eventually contribute to its establishment as a major romance publisher. First, the company began to focus solely on romance fiction as a pragmatic response to the dramatic rationing of printing paper during the war; second, the company began to distribute increasing numbers of books to private lending libraries in the wake of the war, when readers' strained capital resulted in limited book sales (H-MBA 2007c). Romance fiction was an immediate success in this market and Mills & Boon was ideally placed to dominate it. In following decades, the company continued to grow and began to establish a presence in international markets: by the 1950s, for example, Mills & Boon was successfully selling its books in the United States, among other foreign markets. But it was shifts in purchasing habits in the 1960s and 1970s that most profoundly affected the publisher's ultimate success and international influence on the genre: as readers' prosperity improved, lending libraries began to shut down as more and more readers opted to buy, rather than borrow, their books (H-MBA 2006; H-MBA 2007a). Facing a marked decrease in its traditional market, Mills & Boon turned to Harlequin Enterprises Ltd, its highly successful North American distributor, for assistance.

Harlequin Enterprises began as a general publisher in 1949 in Winnipeg, Canada under the comparatively inauspicious name of 'Harlequin Books'; within a decade, Harlequin had changed its name and narrowed its repertoire to focus solely on romantic fiction, which it imported from Mills & Boon (Capelle 1996, 92). However, unlike Mills & Boon, which balanced a moderate commercial interest with a 'gentlemanly nature' – for example, older authors who began to sell fewer and fewer copies of their novels were kept on as 'a kind of charity' or with a paternalistic goodwill towards the authors and the publisher's history (McAleer 1999, 7, 289) – Harlequin was aggressively business minded, applying the mass-market techniques of other industries to publishing to great success. But while the Canadian publisher experienced strong growth in the 1960s, it grew exponentially in the 1970s and 1980s, first through its 1971 acquisition of Mills & Boon and then through Torstar Corporation's 1975 purchase of a controlling interest in the company, only four years after the initial merger of the publishers (McAleer 1999, 284). While both Harlequin and Mills & Boon were influential publishers of the romance genre in the mid-twentieth century – both, for example, were publishing or, in Harlequin's case, distributing in foreign (including foreign-language) markets by the 1960s – the new company, now one organ of the giant Torstar Corporation conglomerate, experienced a period of even greater growth. This growth was based on three strategies: a renewed investigation of ways to market the romance genre; an experimentation with distribution possibilities (for instance, distributing free copies of novels in feminine napkin boxes and an aggressive focus on direct marketing, which continues to this day); and an innovative expansion into international markets throughout Asia-Pacific and Europe (e.g. when the Berlin Wall was destroyed in 1989, Harlequin's staff in West Germany were on hand to distribute 750,000 free romance novels to the newly free women in East Germany, in a typically focused marketing exercise) (McAleer 1999, 284–5; McKay 2008).

The latter is particularly interesting in light of the focus of this paper. Indeed, where both Harlequin and Mills & Boon, like most successful publishers at the time, penetrated international markets by publishing books in their home countries (Canada and England, respectively) and exporting them to foreign markets, the Torstar Corporation facilitated a radical departure from industry norms. Instead of exporting books into international markets, 'separate publishing companies were set up overseas to publish directly in the indigenous language' (McAleer 1999, 284–5). The decision allowed the company to decrease its distribution costs and introduce locally nuanced marketing for its products, with the earliest subsidiaries launched in Australia (1974) and the Netherlands (1975) and other offices following soon after in France, Germany, Italy, Japan, New Zealand and South America. The experiment was a success: by 1981 Harlequin-Mills & Boon had become the largest romance publisher in the world, leading to an unprecedented – and ongoing – diffusion of the mass-market romance genre into the Americas, Australasia, Europe and the United Kingdom. Creed (2003) notes that 'popular romance publishers', but especially Harlequin-Mills & Boon, 'have not been slow to take advantage of the global market' (100).

But with this commercial diffusion, where one national market after another was 'saturated' with the product (see Capelle 1996), came a critical backlash. Indeed, while Harlequin-Mills & Boon novels became the 'base-line definition of romance', it was by no means a celebrated definition (Pearce 2004, 521). Harlequin-Mills & Boon's romance novels were, and are, as commercially popular as they are critically loathed: according to Holmes (2006), the mass-market romance genre was the most derided popular genre in the twentieth century (11–12). Worse, Harlequin-Mills & Boon was the 'least likely' enunciation of the genre to 'achieve any form of legitimacy' in the public imaginary (11–12). Interestingly, both the most ardent detractors and the most ardent supporters explicated the genre in similar ways, revealing troubling public perceptions about the economy of the genre. Specifically, criticisms of the genre (e.g. Harlequin romance novels are 'all the same') and even celebrations of it (e.g. Harlequin romance novels are 'universal', 'universally appealing', etc.) both draw on a universalizing discourse that casts Harlequin-Mills & Boon's enunciation of the mass-market romance genre as overwhelmingly repetitive, with underwhelming (if any) variations between texts (see Thomason 2008; Creed 2003, 100; Pearce 2004, 521; Gallagher 2001, 114). Here, then, mass-market romance is widely conceived to be 'much more tied to formula and convention', much more 'convention-bound', than other popular genres (Gelder 2004, 43; see also Worpole 1984, 33–4).

If genre is partially constituted in public discourse about it, then these perceptions point to a significant public concern about the ongoing viability of the genre: repetition is fundamental to genre, but so is variation. Yet the romance genre's considerable success around the world suggests the exact opposite of any nay-saying public discourse: the genre is both repetitive *and* variant, because that is the necessary structural basis of any successful genre. As John Hartley notes (in O'Sullivan et al. 1994), genre is a dynamic system that is constantly (re)constituted by new additions to it, such that the addition of one genre text automatically shifts and changes, however slightly, the boundaries and makeup of the genre as a whole. Hence, the romance genre's immense international popularity suggests that it has, perhaps more than most genres, continued to find ways to vary its conventions – by adapting and reworking its standards, or its model of writing – to meet the ongoing expectations of its tens of millions of readers worldwide. In other words, while the sites of repetition – the genre's conventions – do stay largely in place in most romance fiction, as in most popular genres, there are nevertheless important sites of variation in each text, irrespective of public sentiment. Even so, these public perceptions are of interest and might point to another, more pointed, criticism about the publisher's negotiation of its creative and economic imperatives. To be precise, the universalizing discourse circulating around Harlequin-Mills & Boon romance fiction might also be an implicit, even tacit, criticism of the publisher's reliance on a mass communication/industrial economic model in operating contexts which are increasingly literate about the value of human creative capital and the affordances of the creative economy. I return to this point below.

Harlequin and/in Australia

Harlequin-Mills & Boon established an office in Sydney, Australia in 1974, one of the first foreign offices to be launched amid Torstar Corporation's ambitious expansion plan. The expansion was key to the publisher's future: as the home market of North America neared capacity, 'foreign outlets enable[d] Harlequin's sales abroad to increase steadily' (Capelle 1996, 92). Foreign income now represents the majority of the publisher's annual sales: in 2007, for example, more than 50% of the publisher's annual sales occurred outside of North America and more than 95% outside of its (Canadian) home market (H-MB 2006). These statistics are perhaps no surprise given the relative success of most of Harlequin's international offices. Harlequin-Mills & Boon's branch office in the Netherlands, for instance, was opened in 1975 – a year after the Sydney office – to test the national market for the potential expansion of the publisher. Despite a conservative launch, which saw only four new novels released into the Dutch market each month, the office was immediately successful. By 1984, 35 titles were published in the Netherlands each month; a decade later, Harlequin-Mills & Boon dominated the national market, accounting for 'roughly 90 percent of the Dutch turnover' (Capelle 1996, 93). Similarly, Harlequin-Mills & Boon conquered the romance market in France so quickly after the late 1970s opening of its Parisian office that local commentators described it as a 'real social phenomenon' (93).

The Sydney office's success has been comparatively modest, but no less effective. After opening in 1974, Harlequin-Mills & Boon Australia now turns over more than A\$20 million per annum and dominates the romance fiction market, the largest segment of mass-market paperback fiction in the country (H-MBA 2007c). Maintaining a 20% market share of national paperback sales and a 90% share of 'women's fiction', the Sydney office publishes around 700 new books each year, a number 'second only to North America', and sells them at a rate of approximately five million books each year, or over 400,000 each month (H-MBA 2007c). In doing so, Harlequin-Mills & Boon Australia is now the largest publisher of paperback fiction in the country, selling more than major competitors like HarperCollins, Penguin, Pan Macmillan, and Random House (H-MBA 2007c, 2008). While neither the market share percentages nor the social celebrations are quite as dramatic as either the Dutch figures or the French commentary, Harlequin-Mills & Boon nevertheless dominates the Australian, and indeed the New Zealand, market.

Of course, the opening of Harlequin-Mills & Boon's Sydney office was by no means the first time the publisher's books had reached Australian shores; Harlequin-Mills & Boon had been exporting its books to the country for a number of years and it was this existing level of product familiarity and success on which the Sydney office was expected to capitalize. The role of the new Sydney branch office was to develop locally nuanced marketing strategies to vigorously distribute 'Mills & Boon' romance fiction throughout the continent.[1] Like the publisher's other branches, Harlequin-Mills & Boon Australia developed and refined several innovations in its ongoing quest to improve its targeting and capture of the national market. But unlike offices in non-English-speaking countries, such as the Netherlands or France, where much of the focus is on literally translating works into the local language and conceptual system (see Capelle 1996), the Australian office has mainly tended to develop innovations around improved service technologies. Three of the office's significant innovations in diffusing the genre in the national market include the following.

In 1998 the Sydney office launched www.romance.net.au, the publisher's first website anywhere in the world. The site featured basic information about the publisher, its authors and its books and, significantly, a shopping feature, enabling readers to purchase a book online and have it delivered directly to them, thus translating the publisher's strength in direct marketing to the online environment (H-MBA 2007b).

In 2000, the office redeveloped www.romance.net.au into www.eHarlequin.com.au, the publisher's current web address, which was refined and adopted by Harlequin-Mills & Boon offices around the world, producing an internationally consistent brand identity and location online (H-MBA 2007b).

In 2008, the office launched an e-book range for PDAs (personal digital assistants), producing a range of new titles as instant digital downloads, making it the fastest way for readers to receive new titles. This innovation also addressed the brand's poor cultural reputation by simultaneously removing the 'embarrassment factor'; by using PDAs to read Harlequin-Mills & Boon content, the books can be read in relative brand 'anonymity' (Newspup 2008).

These innovations, among the largest developed by the Sydney office, are all focused on improving the publisher's targeting and capture of the national market; in other words, these innovations are heteronomous. Heteronomous innovations, as Pierre Bourdieu (1993) said of large-scale productions, are 'favourable to those who dominate the field economically' (40). They court 'mass audiences' by focusing on the 'potentially immediate, broad-based distribution' of 'conventions over originality' (Gelder 2004, 13). Harlequin-Mills & Boon Australia, more than any other publisher currently operating in the country, epitomizes heteronomous production: not only is the publisher synonymous with the ('conventions over originality' of the) mass-market romance genre, but its main innovations over the last decade have also all targeted improved service delivery (or the 'potentially immediate, broad-based distribution' of online shopping and digital downloads). These characteristics – the emphasis on improving profit, decreasing cost and expanding product circulation, alongside the absence of any comparable marketing of the narrative or literary distinctions of mass-market romance – all suggest the publisher's embeddedness within the industrial economy and its one-way mass communication model. Indeed, the publisher's marketing also privileges brand over author name, with any books that are not sold within three months of publication pulped (see Walsh 2008), implying that the mass-market romance genre is a wholly disposable and creatively undervalued one.

However, there are signs that the Sydney office is increasingly embracing the affordances of the creative economy. The most important signal of this was the branch's 2006 introduction of a local commissioning editor, the first in the branch's history. It is hard to overstate the significance of this hiring, in a branch whose approach to the content it distributed had not changed in its three decades of operation. Indeed, this was a crucial shift in the branch's approach to content specifically, and its approach to targeting the national market generally – not least because its parent company had traditionally been quite slow to pursue local content in its international markets. It was only in the 1980s, for example, that the publisher's North American headquarters began to acquire local content (or 'purely American novels') after distributing British content for decades (Capelle 1996, 92). Progress was even slower in Australia: in 2004, Stuart MacDonald, Harlequin-Mills & Boon Australia's (then) sales and marketing director, noted in an interview that the branch's 'trend' was still 'very much towards … drawing content from just one source' – namely from either North America or England (RWAust 2004). Titles were sometimes slightly altered for the Australian market – the 'Desire' series, for example, was renamed from its original British 'Sensual' moniker – but the branch rarely pre-empted 'plans to change' series' names for the 'Australian market', unless 'the original name' was found not to give the branch its 'maximum benefit in Australia' (RWAust 2004). In other words, the branch's approach to the genre was usually retrospective, rather than pre-emptive, and based solely on a product's actual or perceived market success. Outside of these naming issues, content received little attention.

In 2006, however, the company hired its first Australian commissioning editor, signalling its tentative shift away from a branch office operation which distributes products created elsewhere, and towards a creative branch which distributes products it has created. While Australian authors had featured among Harlequin-Mills & Boon's most successful authors for years, until 2006 they had been commissioned through the publisher's North American or British editorial offices. The Sydney office's integration of creation into its existing distribution processes, however, indicated not only its shift towards a creative branch operation but also its concomitant entrée into the creative economy. Unlike the industrial economy, the creative economy 'embraces the entire process from creating the artifact to its marketing, retailing and consumption', the last three of which the publisher had mastered since its 1974 Sydney opening (Rae 2007, 56). Moreover, in the rigid institutional structure of an industrial-era publisher, where national branch offices function as passive distributors of remotely generated content, the recasting of Harlequin-Mills & Boon's Sydney office into a creative branch also points towards the office's intention to interact directly with both its global publishing parent and its national market.

Only two years before the 2008 centenary of Mills & Boon, the Sydney office may have introduced yet another phase in the publisher's extraordinary diffusion of the mass-market romance genre in Australia.

Acknowledgement

This research was supported under the Australian Research Council's Linkage Projects funding scheme (project number LP0777066).

Note

1. At that stage, Australia and the United Kingdom still marketed the books under the anachronistic British branding of 'Mills & Boon'; this practice changed in the 1990s, when the branding was changed to the company's proper title of 'Harlequin-Mills & Boon'.

References

Bourdieu, P. 1993. *The field of cultural production*. New York: Columbia University Press.
Capelle, A. 1996. Harlequin romances in Western Europe: The cultural interactions of romantic literature. In *European readings of American popular culture*, ed. John Dean and Jean-Paul Gabilliet, 91–100. London: Greenwood Press.
Creed, B. 2003. Mills & Boon dot com: The beast in the bedroom. In *Media matrix: Sexing the new reality*, 97–114. Sydney: Allen & Unwin.
Eike, A.M. 1986. An investigation of the market for paperback romance novels. *Journal of Cultural Economics* 10, no. 1: 25–36.
Gallagher, J. 2001. The role of fiction, romance and reading in sustainable development. In *Knowledge, information and development: An African perspective*, ed. C. Stilwell, A. Leach, and S. Burton, 1–22. Pietermaritzburg: School of Human and Social Studies, University of Natal. http://www.infs.ukzn.ac.za/kiad
Gelder, K. 2004. *Popular fiction: The logics and practices of a literary field*. London: Routledge.
Harlequin-Mills & Boon (H-MB). 2006. About Harlequin. www.eHarlequin.com (North America) 2006. http://www.eharlequin.com/articlepage.html;jsessionid=CA186749D21DF48481D9FC2D 2F291CC1?articleId=36&chapter=0 2008
Harlequin-Mills & Boon Australia (H-MBA). 2007a. Fun facts about Harlequin. www.eHarlequin.com.au. http://www.eharlequin.com.au/about_funfacts.shtml
———. 2007b. The eHarlequin.com.au story … so far. www.eHarlequin.com.au. http://www. eharlequin.com.au/about_story.shtml
———. 2007c. The history of Harlequin Mills & Boon. www.eHarlequin.com.au. http://www. eharlequin.com.au/about_history_mb.shtml
———. 2008. Overview, report, & statistics. Unpublished PowerPoint presentation.
Holmes, D. 2006. *Romance and readership in twentieth-century France: Love stories*. Oxford: Oxford University Press.
McAleer, J. 1999. *Passion's fortune: The story of Mills & Boon*. Oxford: Oxford University Press.
McKay, L. 2008. A fine romance with the art of Mills & Boon. *The Age*, 6 September. http://www. theage.com.au/news/books/a-fine-romance-with-the-art-of-mills-amp-boon/2008/09/05/ 1220121527484.html?page=fullpage.
McLean, L. 2008. My speech from Romance Writers of America-San Diego Conference. *Agent Savant*, 19 May. http://www.agentsavant.com/as/index.cfm/2008/5/19/San%20Diego%20R WA%20conference
McLuhan, M. 1967. *Understanding media: The extensions of man*. London: Sphere.

McWilliam, K. 2009a. Genre: Something new based on something familiar. In *Screen media: Analysing film and television*, ed. J. Stadler and Kelly McWilliam, 217–43. Sydney: Allen & Unwin.

———. 2009b. *When Carrie met Sally*. London: I.B. Tauris.

Moran, A., and M. Keane. 2004. Joining the circle. In *Television across Asia: Television industries, programme formats and globalisation*, ed. A. Moran and M. Keane, 197–204. London: RoutledgeCurzon.

Neale, S. 1980. *Genre*. London: BFI Publishing.

Newspup. 2008. Mills & Boon steams up e-books. *Marketing News*, 28 October. http://www.marketingmag.com.au/news/view/mills-amp-boon-steams-up-e-books-804

O'Sullivan, T., J. Hartley, D. Saunders, M. Montgomery, and J. Fiske. 1994. *Key concepts in communication and cultural studies*. London: Routledge.

Orr, K., and M. Stout. 2008. Harlequin romance report 2008. Harlequin. http://www.harlequinromancereport.com

Pearce, L. 2004. Popular romance and its readers. In *A companion to romance: From classical to contemporary*, ed. Corinne Saunders, 521–38. Oxford: Blackwell.

Pettegree, A. 2002. *Europe in the sixteenth century*. Oxford: Blackwell.

Rae, D. 2007. Creative industries in the UK: Cultural diffusion or discontinuity? In *Entrepreneurship in the creative industries: An international perspective*, ed. C. Henry, 54–71. Cheltenham: Edward Elgar.

Regis, P. 2007. *A natural history of the romance novel*. Philadelphia: University of Pennsylvania Press.

Romance Writers of America (RWA). 2007. Romance literature statistics: Industry statistics. http://www.rwanational.org/cs/the_romance_genre/romance_literature_statistics/industry_statistics

Romance Writers of Australia (RWAust). 2004. Hearts talk: Questions asked. Interview with Stuart MacDonald. http://www.romanceaustralia.com/questions.html

Thomason, C. 2008. Why we crave 'happy ever after'. *Manchester Evening News*, 9 July. http://www.manchestereveningnews.co.uk/lifestyle/health_and_beauty/s/1057415_why_we_crave_happy_ever_after

Walsh, J. 2008. Mills & Boon books selling more than ever. *The Courier-Mail*, 28 March. http://www.news.com.au/couriermail/story/0,23739,23439917-5003424,00.html

Worpole, K. 1984. *Reading by numbers: Contemporary publishing and popular fiction*. London: Comedia.

Strategic regionalization in marketing campaigns: Beyond the standardization/glocalization debate

John Sinclair and Rowan Wilken

Cultural adaptation has become a fundamental strategic principle for marketers in the age of globalization. While past decades saw campaigns for products such as Marlboro run on a uniform, 'standardized' basis in every country where they were sold – 'one sight, one sound, one sell' (Mattelart 1991, 55) – marketing has had to learn how to come to terms with the realities of cultural and other differences as it has become ever more globalized.

Standardization has had its advocates, however, and wielded considerable influence amongst global marketers in the 1980s. In particular, a Harvard management guru, Theodore Levitt, proclaimed 'the emergence of global markets for standardized consumer products on a previously unimagined scale' which had overcome 'accustomed differences in national or regional preference' and now required 'the standardization of products, manufacturing, and the institutions of trade and commerce' (1983, 92–3). The leading British advertising agency of the 1980s, Saatchi & Saatchi, helped to build itself into a global corporation by taking up Levitt's doctrine (Mattelart 1991, 48–54).

Saatchi & Saatchi foresaw a world in which there were fewer and fewer agencies servicing fewer and fewer clients, as all industries became more conglomerated in their ownership and management. Indeed, they helped to fulfil their own prophecy to the extent that they fostered a now dominant trend towards 'global alignment'. This is where a global client will have its advertising provided by the same global agency in each and every national market where it does business. Given that such advertising agencies are mostly integrated into 'global groups', huge holding companies which also have market research, public relations, direct marketing and other 'marketing communication disciplines' under their umbrella, the very organization of global advertising, and marketing in general, would seem to facilitate and favour standardization (Sinclair 2006).

Apart from its rhetorical and organizational fit with the nascent global era, standardization was and is seen to have economic advantages: 'the creation of a stronger global international identity through consistent positioning and image across markets over time ... cost reduction through economies of scale in advertising production, sharing of experience and effective use of advertising budget' (Tai 1997, 56–7). That is, corporations have a strong economic disincentive against cultural adaptation and so, as Oscar Wilde said of the dog that could walk on its hind legs, the remarkable thing is not how well it is done, but the fact that it is done at all. Levitt's thesis was much disputed within marketing circles, but more influential in calling it into question were the manifest failures of certain celebrated global campaigns during the 1980s and early 1990s.

Some of these were well documented in the trade press at the time, such as Parker pens; others are more apocryphal, like the warnings circulated in marketing textbooks of instances where slogans or even brand names failed to translate – for instance, that the name Coca-Cola translated into Chinese as 'bite the wax tadpole' (Mooney 2008). As well as such linguistic and cultural differences, including religious strictures and variations in tastes, marketers were also encountering practical differences in national regulatory regimes and distribution systems. This experience turned attention to alternative ways of approaching global marketing, perhaps best represented by Nestlé, which for some time had been pursuing a more localized, or 'multidomestic', strategy of differentially formulating products like its instant coffee in accordance with the peculiarities of various national markets. Standardization has continued to be attractive to global marketers, for the reasons given, but they have had to develop adaptive strategies to cope with market-by-market variations (Herbig 1998).

The legacy of all this has been a kind of continuum evolving in marketing theory and practice between standardization and localization. However, by the beginning of the 1990s, some middle ground was being sought in the concept of 'glocalization', 'one of the main marketing buzzwords' of the time (Robertson 1997, 28). This had its origins in the strategies of Japanese marketers in Asia, notably Sony, which pursued 'global localization' rather than 'global standardization' (Iwabuchi 2002). Just as Roland Robertson took up this unlikely concept in the social theory of globalization to show the fallacious dichotomization of 'homogenization' and 'heterogenization', arguing instead that the tendencies denoted by these terms are 'mutually implicative' (1997, 27), glocalization was embraced in marketing as the practical wisdom of creating the right balance between minding the bottom line of standardization while meeting the demands of localization – more than just a point on a continuum. Indeed, one advocate argues that standardization is a mere abstraction, and all cross-cultural marketing requires some degree of glocalization: it is 'the working arm of standardization' (Herbig 1998, 48).

Adaptation: In search of the right mix, balance and level

Yet there is the crucial question of just what it is that is being glocalized. Marketing textbooks conventionally distinguish between the four 'Ps' in the 'marketing mix': product, place, promotion and price. The product itself is designed with its intended market in mind; it is made available through an appropriate system of distribution; it is promoted in various possible ways, ranging from advertising in the media to more direct contact with prospective consumers such as giveaways; and a suitable price point is set. In his analysis of McDonald's marketing mix, Vignali (2001) adds three more 'Ps': people (staffing and training); process (procedures to be followed); and 'physicals' (the environment of the outlets). The specifics of Mc Donald's as a case study will be returned to later in the article, but the point here is that Vignali distinguishes between those elements that are standardized – indeed McDonaldized (Ritzer 1998) – and those that are adapted to local markets: briefly, for example, procedures remain constant, even if there are local variations in the products and their prices. A different kind of distinction is drawn by Banerjee (1994), who suggests that the physical or functional properties of a product may be kept the same from one country to another, while the brand values – that is, the cultural 'meaning' of the brand – can be varied according to the market. Even in advertising, he suggests, it is possible to use a common advertising strategy and positioning of the product, but vary the 'execution' (102).

A third distinction is made by Tai (1997) in her study of regional advertising strategies in Asia (1997): strategic (that is, main campaign theme, market segmentation and product positioning) versus tactical (creative execution and media placement). Once again, the former tend to be standardized, but the latter localized: she refers to this specifically as an 'adaptation strategy' (58). Finally, and more abstractly, there is the rather old-fashioned distinction between form and content. Robertson, in his discussion of the interpenetration of universalism and particularism, remarks on how, whatever the differences between national societies, '*the form* of their particularities … is very similar (1997, 34). When applied to marketing, this suggests a series of elements which are constant in their presence across markets, but are manifested in quite distinct ways. Thus, although any and all elements within the marketing mix can be varied from one market to another, adaptation consists of holding certain selected elements constant while varying others. In this way, the marketer seeks a balance between the organizational and economic advantages of standardization, and the necessities of responding to cultural and other differences between markets. This is the practical meaning of glocalization.

There is also the issue of exactly who it is that the marketing is directed towards. Echoing Levitt, Saatchi & Saatchi enthused about how global advertising could target 'segments', or similar socio-economic groups in different regions: 'there are probably more social differences between midtown Manhattan and the Bronx … than between midtown Manhattan and the 7th Arrondissement of Paris' (qtd in Mattelart 1991, 52–3). This does seem to be true for certain kinds of products, 'especially those targeting "transnational tribes" of affluent or style-conscious consumers' (Banerjee 1994, 98). These include services that are intrinsically international, like credit cards, and goods that carry international prestige, such as designer brands (Herbig 1998, 49). Yet, as more and more companies – usually based in the United States or Europe – enter the rapidly developing economies of the erstwhile 'Third World', and the BRICs nations in particular (Brazil, Russia, India and China), they are needing to find strategies to cope with the cultural and other barriers that confront not elite but mass market goods and services, such as soft drinks and other FMCG (fast-moving consumer goods) categories, and food franchise operations.

If segments are to be targeted, this suggests that adaptation might have to work at a number of levels: the sub-national as well as the national, and perhaps also the world-regional. In addition, to the extent that adaptation is being made on cultural-linguistic grounds, rather than, for example, to meet national regulations, then geocultural or geolinguistic regions come into consideration – that is, nations or even groups within nations that are not geographically connected (Sinclair 2004). For example, even though contemporary wisdom refers to Asia or even the Asia-Pacific as a region, as Tai observes: 'Asia is really a series of localized markets with their own characteristics, rather than a region' (1997, 49). However, she argues that China (PRC), Taiwan, Hong Kong and Singapore might form a Chinese market – that is, as a geocultural or geolinguistic region.

In this article, we are interested in how global marketers might adapt to the different regions where they do business, not just to the nations within them. That is, if these companies are motivated to minimize adaptation in order to maximize organizational and economic advantages, as has been explained, we would expect that they would attempt to do so on a regional rather than a national or sub-national basis whenever they can. As one marketing academic advises the 'astute' global marketer, they should think globally, act locally and manage regionally (Banerjee 1994, 110). This is what we mean by 'strategic regionalization' in our title. Yet when it comes to looking at such a phenomenon empirically, it becomes apparent that in fact there are not many truly global companies. Most 'global' companies have most of their sales within their own domicile of what Ohmae called the 'triad' of North America, the European Union, and Asia – by the beginning of this decade, of the whole *Fortune 500*, only Coca-Cola had more than 20% of its sales in Asia, while McDonald's had less than 14% (Rugman and Verbeke 2004, 8–9). To take another measure, Procter & Gamble, the world's biggest advertiser, with over 300 FMCG brands, was spending only 20% of its global advertising expenditure in Asia, even though it appeared amongst the top 10 advertisers of most major markets of the region (Endicott 2005). Yet, even if only relatively so, all of these corporations are prominent both globally and in Asia, and are taken here as case studies in the cultural adaptation of marketing.

From Coca-colonization to Coca-localization

According to Theodore Levitt (1983), the Coca-Cola Corporation and its US and global rival PepsiCo are the examples par excellence of his 'globalization of markets' thesis, as both are 'globally standardized products sold everywhere and welcomed by everyone' (93). Certainly, for decades Coca-Cola was distinguished by its highly centralized control of all features of its operation, including marketing. Marketing concepts were developed in the corporation's headquarters in Atlanta, and given minimal local inflections. The fact that by the 1980s the company was deriving most of its revenue from beyond the United States (McQueen 2003, 200) would seem to suggest that such a centralized approach was working well. However, by the end of the 1990s, Coca-Cola was having to deal with a slow but steady decline in the sales of carbonated cola beverages worldwide and in the brand's appeal ('US brands' appeal declines' 2006), and as well with an increasingly difficult struggle to keep pace with PepsiCo, especially in Asia. In response, Coca-Cola began moving away from global, standardized ad campaigns in favour of 'more locally relevant executions' (MacArthur 2000).

This shift in focus was the beginning of what would appear to be a dramatic marketing sea-change towards localization within the company between the late 1990s and 2005. Dispensing with Levitt's globalization of markets approach – which might be summarized as 'think global, act global' – Coke executives adopted a 'think local, act local' strategy (James 2001; McIntyre 2001; Madden, Bidlake, and McKegney 2000). Led by then worldwide CEO Australian-born Douglas Daft, Coca-Cola's new creed became 'there is no such thing as a global consumer: each market is different' (James 2001). This conviction was the result of a growing understanding of cultural, geographic and economic complexities both across and within markets, including market research which found that, even in countries that seem similar, customers can have 'diametrically opposed attitudes' (2001).

In line with this, Coca-Cola restructured its marketing communications operations and designated 13 countries as 'creative hubs' which were to make a pool of ads that could be selected from and adapted to each of the 200 countries where Coke sells its products. One of the first of the ads to be created via this system was produced in Australia and then 'exported' to nine other countries worldwide (McIntyre 2001). At the same time as it was pooling its marketing resources, Coke was also busy developing a range of new product lines for various markets, as it sought to diversify into non-cola drinks such as water products and juices. In most cases, these were in step with its 'think local, act local' strategy and the company's desire to tailor its products to suit local market tastes where necessary (James 2001). A more 'strategic regional' example was a pan-Asian and European 'roll-out', over 2001–2003, of a non-carbonated juice-based drink called Qoo (pronounced 'coo'). Described as a 'regional brand with international potential', Qoo was first launched in Japan in 1999 before entering South Korea, Singapore, China, Thailand, Taiwan and then Germany (Osborne 2001; 'Coca-Cola's Qoo to go to Germany' 2002).

While the adoption of the 'think local, act local' strategy in part contains an admission that 'Coke's local people know their market better than the bigwigs in Atlanta' (McIntyre 2001), it would be misleading to think that this approach represents a total conversion to localization. It is more an amalgamation of global, regional and local strategies, with the emphasis on the sharing of ideas across markets and across brands. Thus, while the company divides itself geographically into three areas – the Americas, the Asia Pacific and Eurasia – brands are not necessarily targeted within these regional boundaries, or even on a geocultural basis. As one Coke executive explained: 'The reality is we know Argentinean and French people do [things] a lot more similarly than Argentineans and Brazilians. That's just a fact in the way they react to the brand and the sort of messages that are right for them' (qtd in McIntyre 2001).

Zandpour and Harich (1996) refer to this approach as 'country clustering' – the grouping of markets which exhibit a preference for similar advertising appeals. Country clustering can be seen as a kind of standardization and is based on the idea that advertisers 'should view standardization not as the transferability of an entire campaign across countries, but as a strategy that makes unified themes, images and even brand names, possible' across countries, even if issues of campaign execution 'still need to be decided at the local subsidiary level' (Sriram and Gopalakrishna 1991, 146). In light of this, Coca Cola's declaration that it is thinking and acting locally is a misnomer. The slogan 'think local, act local' suggests the company is operating at the extreme localization end of the spectrum when this clearly is not the case, as the Argentina–France example illustrates. Evidence indicates that Coca-Cola is working with cross-cultural 'regions' of consumer compatibility, and that its advertising is not locally produced, only locally adapted from a global pool. Nonetheless, Coca-Cola's corporate approach over the past 10 years has acknowledged and attempted to respond to various cultural differences and complexities, and the company has committed to regionalism as part of its global marketing strategy – just no more than it had to.

Fries with that? I'm Lovin' It

In 2004, McDonald's Global Chief Marketing Officer Larry Light caused a stir in advertising circles by declaring that 'the days of mass marketing are over' (qtd in Cardona 2004). McDonald's was said to have 'ditched traditional brand-positioning marketing' in favour of an alternative approach which Light referred to variously as creating 'brand journalism' or a 'brand chronicle' – that is to say, taking a narrative approach that 'seeks to tell as many different stories in as many different ways as it takes to reach McDonald's 47 million consumers in 119 countries' (2004). For Light, McDonald's 'I'm Lovin' It' campaign has been a crucial element in this 'brand journalism' approach, and this campaign has, by his own admission, 'reinvented a brand that had lost its way' (2004). The origins of the slogan behind this campaign, launched in 2003, can shed light on McDonald's overall global marketing strategy and its position on the standardization–localization debate.

By 2004, McDonald's had moved away from the company's traditional approach of having the agency office closest to headquarters taking the lead in creating its global campaigns (Cardona 2004). In a move not unlike the 'country cluster' approach adopted by Coca-Cola, McDonald's drew more on its agencies around the world to produce creative ideas which it could then review for selection. In the case of 'I'm Lovin' It', this slogan emerged from a lesser known European agency, DDB's Heye & Partner, Hunterhaching, Germany. Once selected, however, the transformation from idea into a fully fledged campaign was done centrally, with creative work handled on a global basis, but with local agencies given the opportunity to adapt the campaign to suit each national market. An example cited by Light as to how this works in practice is through the creation of a global television commercial (TVC) 'template' with 'green-screen segments' for local agencies to 'insert local touches' (Cardona 2004). So, for instance, in adapting North American TVCs based around the 'I'm Lovin' It' theme for mainland China, the American singer Justin Timberlake was replaced by Wang Leehom, a popular local singer performing a Mandarin version of the campaign jingle (Madden 2003). This strategy was repeated with other minor variations in each of McDonald's Asian markets (Liu 2003; Madden 2003).

Just as McDonald's picked up on the initiative of a local advertising agency and then globalized it in the case of the 'I'm Lovin' It' campaign, the corporation has shown itself willing to expand upon local innovation in other aspects of marketing. In 1993, the 'McCafé' – a European-style coffee shop within a regular McDonald's outlet – was introduced by McDonald's Australian management to a 'problem store' in Melbourne, possibly at Victoria Market, where there is a thriving coffee culture (Jackman 2004). By 2001, it had been introduced to 17 countries, including the United States ('McDonald's opens first McCafé in the US' 2001), but can also be found in Asian countries such as Singapore and Japan, with a special menu ('McDonald's McCafé coffee shops debut in Japan' 2007). In other words, McDonald's has taken an innovation which was prompted by local tastes and style in one country and, as with the TVCs, introduced it to other countries in a glocalized form.

The overall process evident in these instances – developing a single global campaign drawn from local sources and redesigned for local adaptation – is indicative of McDonald's philosophy of 'think global, act local', to invoke the globalization cliché of the 1990s. This approach informs all facets of the company's global business operations, as well as its worldwide marketing communications strategy. All of McDonald's operations are in some way structured around global–local tensions. As noted above, certain facets of its global operations are more readily standardized than others, such as food preparation protocols, point of purchase, signage and design. In some cases, the standardized business practices have their own modes of adaptation to the cultural specificities of emerging markets. For instance, McDonald's has long been committed to utilizing local staff and promoting from within, and this has the benefit of developing managers 'who understand both the corporate and local cultures' (Vignali 2001, 107; Watson 1997, 12–14). In other words, McDonald's corporate training program is considered a standardized strategy, but one that has distinctly 'local' outcomes and advantages.

A further standard/local strategy that has been pursued throughout McDonald's history and across all its international markets is the company's appeal to children. McDonald's has vigorously targeted children, not only through television advertising but also via various 'below the line' promotional strategies such tie-in deals, in-store birthday parties, and the provision of 'safe' on-site play equipment (Watson 1997, 19). In this way, regardless of the nation in which they grew up, people come to think of McDonald's as a part of the local culture of their own childhood, so that McDonald's is 'no longer perceived as a foreign enterprise' (37).

Other aspects of McDonald's operations that may require substantial adaptation to meet local conditions include pricing structures and menu options. Pricing structures are tailored to meet the various socio-economic capacities of consumers in each market, and what is considered the most acceptable price is always measured against the company's local competitors (Vignali 2001, 101–3). The global variability of prices for a standard Big Mac hamburger has in fact become an informal measure of purchasing power parity (The Big Mac Index 2008). As to standardizing menus, although there are substantial cost savings to be gained by doing this, McDonald's has discovered that localizing its menus is often crucial to its long-term success, especially to gain cultural acceptance as part of the wider market 'foodscape' in emerging markets. This is one of the most interesting aspects of adaptation in the McDonald's case, for although McDonald's has actively localized its menu options, in one key feature it has maintained a consistent and highly standardized approach.

McDonald's regularly tailors menus to meet various cultural and religious laws and customs, as well as taste preferences in different countries. There are numerous examples in the marketing literature: ginger egg tarts in Taiwan, rice burgers in Hong Kong, taro pies in China, teriyaki burgers in Japan, grilled salmon sandwiches in Norway and poached egg burgers in Uruguay (Watson 1997, 23–4; White 2006; Desker Shaw 2006; Fowler and Setoodeh 2004; Vignali 2001, 99; 'McDonald's puts fish on the menu' 2003). Even so, the 'keystone' of McDonald's 'winning combination' is also the most standardized item on the menu: its 'French fries' (Watson 1997, 24–5). These are 'everpresent and consumed [worldwide] with great gusto' by all McDonald's customers, 'irrespective of their religious beliefs or political stance' (Vignali 2001, 100). Once again, we see that although there is a necessary willingness to adapt some features, others have to be held constant: although the Chinese, for example, want to find menu choices which are palatable to them, they do not go to McDonald's to eat Chinese food, and evidently enjoy the dubious attractions of the iconic fries as much as anyone else in the world.

Finally, there are several indications from the literature over the past decade that McDonald's is becoming increasingly interested in adapting standardized marketing messages at a *regional* rather than local level. For instance, in 1998 McDonald's ran its first pan-European campaign to coincide with the France '98 World Cup football finals (Euronews 1998). More recently, in 2005, the company announced its 'first pan-Asia initiative': the 'Prosperity Program' (Madden 2005; Desker Shaw 2005). This was a program based on its 'Prosperity Burger', and employed the promise of good luck as an 'insight that cuts across borders of nine very diverse markets' (Madden 2005). Further pan-Asian marketing programs followed in 2006 (Hargrave-Silk 2006). We have seen that McDonald's strives to keep certain elements of its global operations constant, while it adjusts other elements when and where required. This has not changed over the past 10 years, but what has changed is the company's apparent interest in doing this on a regional basis.

Procter & Gamble: Flexible organization and total control

As the global manufacturer and marketer of hundreds of brands of FMCGs, Procter & Gamble (P&G) is perennially the world's biggest advertiser, and the trade press has enthused over its transformation from an "'old economy" dinosaur' to a state-of-the-art marketer in the new century ('Well-balanced plan' 2005). However, while Coca-Cola has been experimenting with greater localization of its operations, P&G has undergone a major operational restructure, dubbed 'Organization 2005' (or O-2005), which has had the effect of 'further centraliz[ing] an organization already more centralized than its peers' (Neff 1998). Initiated in 1997, the O-2005 review was 'designed to balance global and local advantages and considerations' (Dyer, Dalzell, and Olegario 2004, 308). Prior to the restructure, P&G was organized into four regions – North America, Latin America, Asia, and Europe/Middle East/Africa – with 'category managers' functioning within each region (Neff 1998). The new corporate structure created a 'matrix' with two axes: along one, P&G grouped its brand management groups into seven global business units (GPUs), while along the other were 'eight geographically based market organizations (MDOs)' (Dyer, Dalzell, and Olegario 2004, 294–5). According to historians of P&G, this elaborate structure was intended to 'balance global advantages of scale with the particular demands of local circumstances', thus: 'The GBUs would develop and manage strong local brands', by developing largely standardized global products, while the MDOs would be more responsible for the execution of marketing strategies, including through the adaptation of standardized advertising, and to a lesser extent the products themselves, to local conditions (Dyer, Dalzell, and Olegario 2004, 294–5).

The O-2005 plan was to simplify the structure and chain of command of the company's worldwide operations (Neff 1998). To this end, P&G also set about streamlining its agency relationships through even tighter global alignment. For example, in the late 1990s the firm consolidated its accounts globally with its 'favoured four' agencies: Saatchi & Saatchi, D'Arcy Masius Benton & Bowles, Grey Advertising and Leo Burnett Co. (Snyder and Neff 1999). Then, in 2003, P&G approached its two main agency holding companies, Publicis Groupe and Grey Global Group (now part of WPP), asking that they 'pitch plans to consolidate or re-organize its global retail marketing effort' (Neff 2003a). On the other hand, P&G has also actively set out to 'decentralize' (or 'localize') media buying – that is, the strategic purchase of advertising time and space in the media, including the Internet – seeing this as more of a nation-by-nation function of advertising (Neff 2003b).

These manoeuvres aside, the overall company philosophy on the issue of standardization and localization (adaptation) remains unequivocal. As P&G's former Global Marketing Manager, Jim Stengel, put it, the 'organization is flexible enough to work between the extremes of standardized global brands and total local control' (Neff 2002). According to Stengel, global branding has to be thought of as a continuum 'that has absolute standardization on one end, and total local adaptation on another end' (qtd in Neff 2002). Global brands, he suggests, are flexible enough always to be somewhere on that continuum. Consistent with the discussion earlier in this article about holding some factors constant while adapting others, Stengel distinguishes between decisions that have to be made on 'hard points' (consistencies across markets, such as manufacturing equipment) and 'soft points' (elements that can be changed, like brand names, colours and patterns) (qtd in Neff 2002). A good example of this is laundry detergent, where the company's long-running Tide line of products has undergone minor modifications to brand naming and packaging, to be advertised under the Ariel label in Latin America and Europe, with considerable success and remarkable consistency with the brand's North American equivalent. However, by its own admission, P&G has sometimes gone too far down the standardization path. A salutary illustration of this was the company's disastrous decision to change the name of Escudo, a bar soap strong in Mexico, to its North American brand name of Safeguard: sales plummeted until the product name was switched back to Escudo, which saw volume sales return again (Neff 2002).

To summarize, the key development for P&G over the past 10 years has occurred at the organizational level, with the company realigning its worldwide operations to achieve greater synergy between its global brand categories and its regionally oriented market development organizations. Despite these structural upheavals, however, the company's approach to branding, product development and advertising has remained relatively stable. All are ultimately 'glocal' in orientation, with brands, products and campaigns developed and set centrally and then adapted to suit individual markets.

Conclusion: Glocalization and strategic regionalization

On the basis of the cases examined here, Theodore Levitt's prophecy of the globalization of markets and universal standardization has not in fact eventuated, but, by the same token, neither does universal localization exist. Rather, the dominant global marketing approach is that of 'glocalization' – an amalgam of global strategy and local adaptation. Within this model, the specific approach that each corporation takes can shift depending on which aspect of its overall operations is involved. Thus, organization, product and advertising can and often are globally aligned or locally adapted to differing degrees depending on the company, the particular point in time and other circumstances. On the question of the degree to which glocalization is being practised on a regional basis, the approach of the global corporations under discussion here suggests what we are calling strategic regionalism, where organizational structure, ad creation and marketing strategies have, over the past decade, been realigned to varying degrees and in different ways around the concept of the global region. Even the notion of 'country clustering' can be considered a regional strategy of sorts, albeit one that cuts across geographical regions, working cross-culturally to identify, group and target markets with cultural affinities, and compatible consumer behaviour and tastes. Regionalization appears to be of continuing interest to global marketers, and, as this study has also shown, there are many different forms of strategic regionalism that may be mobilized. These range from the concentration of corporate operations within certain geographical regions (P&G), to marketing efforts tailored to geographical regions (McDonald's), geolinguistic regions (P&G) and supra-national 'geocultural regions' based on 'clustering' of cultural and other compatibilities (Coca-Cola).

Acknowledgement

This paper is an output from a program of research under Australian Research Council Discovery Grant – Project DP0556419, 'Globalisation and the Media in Australia', funded 2005–09. The authors gratefully acknowledge the ARC's financial support.

References

Banerjee, A. 1994. Transnational advertising development and management: An account planning approach and a process framework. *International Journal of Advertising* 13, no. 2: 95–124.

Cardona, M. 2004. Mass marketing meets its maker. *Advertising Age*, 21 June: 1.

Coca-Cola's Qoo to go to Germany. 2002. *Advertising Age*, 16 December: 12.

Desker Shaw, S. 2005. McDonald's runs first pan-Asia brand drive. *Media*, 14 January: 11A.

———. 2006. McDonald's links push to HK mood. *Media*, 7 April: 8.

Dyer, D., F. Dalzell, and R. Olegario. 2004. *Rising tide: Lessons from 165 years of brand building at Procter & Gamble*. Boston: Harvard Business School Press.

Endicott, R.C. 2005. 10th annual global marketing. *Advertising Age*, 16 December: 1–53.

Euronews. 1998. *Advertising Age International*, 8 June: 9.

Fowler, G.A., and R. Setoodeh. 2004. A question of taste. *Far Eastern Economic Review*, 12 August: 32.

Hargrave-Silk, A. 2006. McDonald's loyalty scheme to go regional. *Media*, 27 January: 14.

Herbig, P.A. 1998. *Handbook of cross-cultural marketing*. New York: Haworth Press.

Iwabuchi, K. 2002. *Recentering globalization*. Durham, NC: Duke University Press.

Jackman, C. 2004. Burger King. *The Weekend Australian Magazine*, 23–24 October: 20–5.

James, D. 2001. Local Coke. *BRW*, 20 September: 70.

Levitt, T. 1983. The globalization of markets. *Harvard Business Review*, May–June: 92–102.

Liu, C. 2003. McDonald's brings its 'lovin" to China. *Media*, 3 October: 4.

MacArthur, K. 2000. Coca-Cola Light employs local edge. *Advertising Age*, 21 August: 18.

Madden, N. 2003. Spotlight. *Advertising Age*, 6 October: 18.

———. 2005. Spotlight. *Advertising Age*, 31 January: 16.

Madden, N., S. Bidlake, and M. McKegney. 2000. Coke loosens ad reins. *Advertising Age International*, April: 1–3.

Mattelart, A. 1991. *Advertising international: The privatisation of public space*. London: Routledge.

McDonald's McCafe coffee shops debut in Japan. 2007. *Japan Today*,. http://archive.japantoday.com

McDonald's opens first McCafé in US. 2001. http://www.enterpreneur.com

McDonald's puts fish on the menu in Singapore sell. 2003. *Media*, 31 October: 6.

McIntyre, P. 2001. Bursting the global bubble. *The Australian*, 2 August: M9.

McQueen, H. 2003. *The essence of capitalism*. Sydney: Sceptre.

Mooney, P. 2008. Bite the wax tadpole? The Coca-Cola Company. http://www.coca-colaconversations.com

Neff, J. 1998. P&G revamp to create global category groups. *Advertising Age*, 22 June: 2.

———. 2002. International advertising: P&G flexes muscle for global branding. *Advertising Age*, 3 June: 53.

———. 2003a. P&G to slash roster for $3B retail business. *Advertising Age*, 7 July: 1.

———. 2003b. $3.6 billion outlay: Glock to guide P&G global ad spending. *Advertising Age*, 1 September: 3.

Osborne, M. 2001. Qoo is coup for Coke in Asia. *Advertising Age*, 8 October: 16.

Ritzer, G. 1998. *The McDonaldization thesis: Explorations and extensions*. London: Sage.

Robertson, R. 1997. Glocalization: Time-space and homogeneity – heterogeneity. In *Global modernities*, ed. M. Featherstone, S. Lash, and R. Robertson, 25–44. London: Sage.

Rugman, A.M., and A. Verbeke. 2004. A perspective on regional and global strategies of multinational enterprises. *Journal of International Business Studies* 35: 3–18.

Sinclair, J. 2004. Globalization, supranational institutions, and media. In *The Sage handbook of media studies*, ed. John D.H. Downing, 65–82. London: Sage.

———. 2006. Globalisation trends in Australia's advertising industry. *Media International Australia* 119: 112–23.

Snyder, B., and J. Neff. 1999. P&G eases conflict policy. *Advertising Age*, 1 February: 1.

Sriram, V., and P. Gopalakrishna. 1991. Can advertising be standardized among similar countries? A cluster-based analysis. *International Journal of Advertising* 10, no. 2: 137–49.

Tai, S.H.C. 1997. Advertising in Asia: Localize or regionalize? *International Journal of Advertising* 16: 48–61.

The Big Mac Index. 2008. http://www.economist.com

US brands' appeal declines. 2006. *Advertising Age*, 27 February: 26.

Vignali, C. 2001. McDonald's: 'Think global, act local' – the marketing mix. *British Food Journal* 103, no. 2: 97–111.

Watson, J.L. 1997. Introduction: Transnationalism, localization, and fast foods in East Asia. In *Golden arches east: McDonald's in East Asia*, ed. J.L. Watson, 1–38. Stanford: Stanford University Press.

Well-balanced plan allows P&G to soar. 2005. *Advertising Age*, 12 December: S2.

White, A. 2006. Exotic flavour makes McDonald's shrimp spot a winner. *Media*, 7 April: 22.

Zandpour, F., and K.R. Harich. 1996. Think and feel country clusters: A new approach to international advertising standardization. *International Journal of Advertising*, 15: 325–44.

Recombinant Broadway

Jonathan Burston

Prologue

A brief scan of recent *Variety* items sets the scene. Underneath a banner front-page headline shouting 'Hollywood's Musical Mania!' Tatiana Siegel reports on the relief Los Angeles executives are feeling as screen musicals are cycling away from mass indifference and, just on time for a recession, becoming popular again. 'It's beginning to look a lot like the 1930s', Siegel writes. 'The economy is in the toilet, and Hollywood studios are filling their pipelines with upbeat dance films, particularly teen hoofers' (Siegel 2008, 1).

Siegel goes on to list the musical (or 'tuner') rosters of players such as Disney and Paramount, Columbia and MGM (retreads of 1980s chestnuts *Fame* and *Footloose* are coming soon to a cinema near you). But smaller outfits are also jumping on the bandwagon. Beneath the headline 'Harvey Enters from Wings – Exec's Hit Movies Move Center Stage as Tuners', Gordon Cox (2008a) reminds us that, at the other end of the continental flyover, Broadway is overjoyed at Hollywood's newfound interest in the musical form as figures like Miramax legend Harvey Weinstein busily turn their film properties into stage musicals. Weinstein is preparing stage musical treatments for his *Shakespeare in Love*, *Finding Neverland*, *Cinema Paradiso* and *Chocolat* – probably in some instances only to morph them once more into screen musicals in the future. When this happens, he will be following a successful new profit-maximization strategy for his cinematic-cum-theatrical properties that has already been proven by John Waters'

Hairspray (2002),[1] which was a small film with music, then a mainstream stage musical, then a blockbuster movie musical hit, and Mel Brooks' *The Producers* (2001), which followed the same triple play, if not yielding quite as gratifying a box office result in its last incarnation.

In fact, Weinstein has been in the Broadway producing business for 10 years now. Like a number of other smaller Hollywood producers who have been making their mark on the New York theatre scene over the last decade or so, he has decided to produce musical and non-musical fare on and off Broadway in the hope that he can replicate the good old days of the mid-twentieth century, when successful Broadway properties were regularly adapted for the big screen.

But here we encounter an intriguing departure from the way things were done during Broadway's self-styled Golden Age. Historically, when stage productions made the transfer to movie properties, Broadway and Hollywood were in competition with one another even as they did business together. Broadway producers used to guard the rights to their stage properties jealously, wary of ceding anything to their Hollywood counterparts until box office receipts for their Broadway productions were clearly waning. Conventional thinking argued that releasing a *Sound of Music* movie while the show (1959) was still going strong on the Rialto (aka Broadway) would eat into revenues for the stage production.

This isn't the thinking any more. Producers today are too accustomed to reading *Variety* headlines like 'Pic Lays its Love on "Mamma": Adaptation a Super Trouper for Rialto, Road' (Cox 2008b)[2] to follow the old rules. Indeed, an entirely new set of rules now seems to apply. So below the headline, *Mamma Mia!* producer Judy Craymer told Gordon Cox:

> I waited on making the movie until I felt it would have maximum impact on the show ... It was important to me that the movie had a major worldwide distribution, with all the marketing that goes with that, to support the stage productions around the world. (Cox 2008b, 36)

Indeed, ever since 2002 when the arrival of the movie version of *Chicago* (1977) resuscitated the flagging box office fortunes of a long-running stage revival, Broadway producers have viewed film adaptations of their properties as helpers, not hindrances, when their productions begin to display signs of box office wear and tear. And if the Broadway tallies for *Mamma Mia!* (1999) provided any indication of the kind of 'B.O. bump' the show's productions received in Las Vegas, London, and four other non-US locations, Craymer made the right choice. Subsequent to the movie's premier, Broadway wraps – or future sales – alone were up 'by an impressive $300,000 per week' (Cox 2008b, 39).

In fact, live-theatrical producers all over the world are newly in love with the cinema. 'South Korea's film industry may be struggling though a painful downturn', writes *Variety*'s Han Sunhee (2008), 'but its stage musicals biz is on the upswing. And a flood of new tuners adapted from local movies is accelerating that growth.' So popular are these adaptations that the national media have even given the genre a new name – 'movicals'. Until recently, the Korean market for stage musicals ranged from moribund to modest. But according to industry estimates, in 2006 the market for stage musicals 'passed the $200 million mark ... making it one of the most profitable forms of popular entertainment in Korea' (Sunhee 2008, 31).

Meanwhile, back on the musical's native soil, Broadway's newfound love for Hollywood has hardly gone unrequited. And it has hardly been limited to Harvey Weinstein. *Shrek – The Musical* (2008), *Mary Poppins* (2004), *Legally Blonde* (2007),

Xanadu (2007), *The Color Purple* (2005), *Dirty Rotten Scoundrels* (2005) and *9 to 5* (2008) have lately joined a roster of stage musicals whose birthplace is Burbank, not Broadway. This is a roster that has been growing conspicuously over the last 10 years. Stage musical versions of *Saturday Night Fever* (1998), *The Producers* and *Hairspray* were gambles that Broadway was prepared to take, thanks in no small part to Disney's notable success in the mid-1990s with it first stage production, *Beauty and the Beast* (1994). *Beauty*'s eponymous box office was quickly matched, then overtaken, by *The Lion King* (1997). Since then, *Tarzan* (2006) and *The Little Mermaid* (2008), among others, have joined the parade.

Circumstances such as these leave some seasoned observers of musical theatre scratching their heads. What is going on here? How could the stage musical's fortunes have changed so dramatically? What makes the musical such a good business bet all of a sudden, even in unlikely new markets? Why are these endlessly recursive adaptations and readaptations suddenly so popular, and should we be happy about it?

Answers to these questions are discoverable in Broadway's own globalization tale. Like the roots of other globalization tales, Broadway's are buried further back in the past than current popular discourse often suggests (see Moran 1998). But things did begin to change rather markedly in the early 1980s – as was true for the cultural industries in general – and these changes continue to reverberate on the field of live-theatrical production. Like so many other transition stories inside our brave new world of recombinant entertainment, the story of today's Recombinant Broadway is intimately wrapped up in the changing motivations and capacities of capital, the recent progress of digital technologies, and the constant movement between Fordist and 'post-Fordist' moments of production in neo-liberal times.

Act 1: *Cats*, *Phantom* and the onset of theatrical Fordism

Back in the early 1980s, famed British theatrical producer Cameron Mackintosh and superstar composer Andrew Lloyd Webber found themselves in a quandary. Their new musical, *Cats* (1981), had become astonishingly popular. Productions in London's West End, on Broadway, and tours in North America and Europe couldn't begin to meet audience demand. What they needed to do, they decided, was to figure out how to 'roll out' new productions into new markets quickly – more quickly than had ever been done before. What was more, they determined, most of these shows wouldn't be touring productions. Rather, they would 'sit down' in emerging major markets like Hamburg, Tokyo and Toronto for as long as there were adequate audience numbers, which often turned out to be years and years. Crucially, Lloyd Webber and Mackintosh also concluded that each show should look, feel and sound *exactly* like the originating productions in the West End if the producers were to maximize the new levels of profitability that appeared to be out there – levels that had only been dreamt of when a show's 'rollouts' had been limited to two theatrical capitals on either side of the Atlantic and their associated tours.

They thus set out to learn how to 'clone a show' – terminology deployed by Lloyd Webber himself in describing the procedures that quickly evolved to mount *Cats*, and then *Phantom of the Opera* (1986) and many of his subsequent 'megamusicals' (Lister 1995). Remounts of these giant new shows, and of similarly mega-sized Mackintosh properties like *Les Misérables* (1985) and *Miss Saigon* (1989), didn't work like they did for shows of the mid-twentieth century when West End productions of Broadway hits often had different designers or even different directors than the ones who had been attached to the original New York production. Procedures for remounting megamusicals owed more to

Fordist logics of the production line than to the craft-based models of reproduction that had preceded them. Throughout the 1980s and 1990s, and into the new century, actors, musicians and others working on megamusicals have complained of new restrictions on their creative autonomy as new tiers of globe-trotting artistic staff 'directed' performers on matters pertaining to blocking, gesture and interpretation with new and profoundly alienating levels of precision. In shows deploying the megamusical reproduction model, actors in far-flung locations are told – sometimes explicitly, often by implication – to reproduce performances that were identical in every way to those first crafted in London or New York. Managerial directives such as these have been buttressed by licensing agreements that specify how the re-mount in question should replicate the original production precisely, from the sets and costumes down to displays for show merchandise in the lobby (Burston 2000; Walsh and Platt 2003; Russell 2007).

Megamusical runs can be long and they often afford actors higher degrees of job security than anything they have enjoyed previously. But many have chafed nonetheless under their new working conditions, complaining regularly about the 'assembly line' nature of megamusical work, about becoming acting 'machines' and about having to work 'McTheatre jobs' (Burston 2000; Russell 2007). Indeed, jobs such as these are more or less unprecedented in theatrical history. Barring various experiments undertaken last century by Vsevelod Meyerhold, Edward Gordon Craig and other avant-garde directors, theatrical performances have been understood to be 'theatrical', after all, in part because they are undertaken by different people in different times and places, and therefore not identical to one another. Rather than being the pre-moulded products of industrial labour, they have been understood as 'one-off' productions undertaken by guilds of artists and craftsmen.

Each of these craftsmen, as Diderot put it in his famous treatise on acting, might 'gather strength with the new observations he will make from time to time. He will increase or moderate his effects, and you will be more and more pleased with him' (Diderot 1883, 8). The *philosophe* here speaks aloud a shared understanding among Western theatre audiences going as far back as Aeschylus: audience satisfaction tends to derive in no small part from the discernment of interpretive difference. The notable, if sometimes unwitting, shunning of mainstream theatrical convention described here, however, has been largely invisible to vast numbers of megamusical audience members. Many are unaware of the new conditions of production under which actors labour. And many have been drawn to the theatre for the first time by production values that are so extravagant (falling chandeliers in *Phantom*, descending helicopters in *Saigon*, flying beasts in *Beauty*) that matters of performance and interpretation seemed to become a second-order concern anyway – icing on the spectacular cake (Burston 2000; Walsh and Platt 2003).

From the very beginning, the Fordist qualities of megamusical production were equally visible in the financial register as Lloyd Webber, Mackintosh and newer giant producers like Toronto's Livent began emulating Hollywood-sized global marketing campaigns for their shows. They also began deploying strategies of vertical integration, hoping to build theatrical empires that would control every moment of the production process. Mid-size theatrical producers began embarking on joint production ventures in order to bulk up and get big enough to stay competitive in this new world of theatrical producing where, thanks to megamuscials' enormous new budgets, barriers to entry were becoming formidable indeed (Burston 2000).

Likewise from the onset of theatrical Fordism, the new cult of verisimilitude thrust upon megamusical actors was reinforced by designs for lighting and sets that aimed for correspondingly identical results from production to production. New digital advances provided new, gargantuan degrees of proportion, automation and technical complexity to each kind of designer. Significantly, both usually deployed these new advances to promote a distinctly non-theatrical, quasi-cinematic realism to the stage – a particular kind of visual 'seamlessness', as one prominent producer described it in interview, that aimed to erase all visible evidence of wires, pulleys and other cues to the existence of a long-standing theatrical tension between the actual and the contrived. A newly digitized soundscape accompanied this new visual code, one that reproduced a show's music in the theatre as if it was being played on a CD. This has since become a new sonic norm for stage musicals everywhere – one where the overuse of microphones, speakers and sound boards can rob singers of much of their own vocal agency. The audio mixer's new power to control levels of volume and qualities of timbre often creates a weird homogeneity among voices – one, moreover, that is entirely unnecessary, given the capacities of contemporary audio technology (Burston 1998). In humiliating addition, it is often difficult to determine precisely who on stage is speaking or singing at any given time (Canby 1995). In sum, Mackintosh and Lloyd Webber had devised a comprehensive set of strategies to lock down live performance in the service of mass reproduction and its associated revenue potential.

Act 2: The Disney moment

In the 1990s, bigger players – most especially Disney – having noted well the string of global successes that Lloyd Webber and Mackintosh had lately recorded under a new theatrical Fordism, followed the British producers into the field of live-theatrical production. Because of its capacities as transnational media corporation, Disney succeeded in radically increasing the potential profitability of the production model first pioneered by the British producers. Former Disney studio chairman Joe Roth explained how in 1996, by talking about the release of one of his company's animated features:

> A major studio spends to stimulate all of the revenue streams, from merchandising to video to theme parks. Look at *The Hunchback of Notre Dame*. It will gross $300 million worldwide, but when you look at all revenue streams that number more than doubles. (Qtd in Bart 1996, 56)

By 1999, *Hunchback*'s revenue streams included a legit production in Berlin, though it disappeared rather quickly. Prior to this, however, 1997 saw the opening of Disney Theatrical's second major Broadway property, the massively successful *Lion King*, whose business plan appeared to be led by the promise of ancillary revenues in the event of disappointing box office. A 1998 *Variety* piece, 'B'way Rules Rewritten to Heed "Lion's" Roar', revealed a crucial new development in the history of theatrical financing. The increasing centrality of synergy strategies among transnational producers – all of whom can exploit theatrical properties across other media and merchandize – meant that theatrical recoupment was no longer as serious a concern as it had been for producers of the old school.

Although it did recover its costs in spectacularly short order, *Variety* reported on Roth's opinion that it didn't really matter whether the enormously expensive *Lion King* ever, in fact, recouped on Broadway: '"More than anything, it's the perception." He adds that, as with film, the answers are in the foreign and the ancillary streams of video, merchandizing and soundtracks' (Cox and Evans 1998, 78). As a second executive explained (78), Disney's *Lion King* strategy did not require the Broadway production to turn a profit. Rather, 'if you break even on Broadway, you're feeling OK. You set up Broadway as the marketing point for the rest of the world.' This applies both to ancillary products and to future theatrical productions. In January 2007, Disney Theatrical president Thomas Schumacher told the *New York Times* that, although the company's new *Tarzan* had failed to connect with either audiences or critics in the several months since its opening, the show was covering its costs and therefore it made sense to keep it open at the Richard Rogers Theater. Two productions of *Tarzan* were being planned for Europe, and maintaining *Tarzan* on Broadway was clearly good for the non-Broadway *Tarzan* franchise (Robertson 2007).

Synergy is a term whose currency rises and falls almost seasonally. Yet, despite regular breaches between synergy's theory and its practice, the economies and efficiencies of synergy are often understood to be more prevalent now than they were 30 years ago – thanks to the onset of digital convergence. According to entertainment industry maven Michael J. Wolf: 'Content is becoming a very liquid asset' in a world whose entertainment products are all translatable to binary code:

> Whether you experience *Men in Black* in a theater, on video, on [DVD], on HDTV, on a CD-ROM, in a video game, in a soundtrack, or over the Internet, the digital content remains the same. The same digital language captures Will Smith's performance of the title track and *Entertainment Weekly*'s story on the movie's special effects (which were also created in a digital language). Digital technology is thus uncoupling entertainment products from any specific medium and making them portable across multiple platforms. In a world where most content moves at the speed of light, there is an even greater need to be available in every medium and every platform in order for any new product to have a chance against the competitor's blockbuster multiplatform effort. (Wolf 1999, 92)

Thus, with the arrival of transnational media corporations on the field of live-theatrical production, the stage becomes another platform for their efforts to fully exploit content which, as we are reminded ad nauseam in the trade press, is king in the digital era. 'Content' is what gets loaded on to Wolf's aforementioned platforms, such as DVDs, Playstations, or YouTube.com – platforms which, as Wolf rightly suggests, are increasingly multiple in number. And content, in its live-theatrical incarnation, is any given intellectual property on the market, be it an animated feature (Disney), a movie like *The Full Monty* (News Corporation/Fox Searchlight) or a book with screenplay potential like *Wicked* (NBC Universal), whose profit potential a transnational media corporation wishes to maximize. As John Scher of the now defunct PolyGram put it when PolyGram owned a 30% stake in Lloyd Webber's Really Useful Group: 'We are an entertainment company, and we are looking for product to feed our systems' (Gubernick 1994).

Since then, many more producers have concluded that synergy strategies common to the marketing of other entertainment properties can now be extended into the live-theatrical field very profitably indeed. Some of these newer producers aren't even giant media corporations, suggesting new degrees of market maturation. But like *Mamma Mia!*'s Judy Craymer, they've been watching developments on the street for some time and are now tearing pages from the Disney playbook in order to vault their own properties into the game.

Act 3: Recombinant Broadway

Broadway is often called 'the fabulous invalid', forever teetering as it appears to be on the brink of financial collapse. Historically, industry fundamentals including high output costs and fickle audiences have contributed to investment conditions that rarely guarantee returns. Being a Broadway producer has always been a risky business – so risky that by the famously lean 1970s, as live theatre's star continued to fade and those of the cinema, television and newer forms of popular music shone ever more brightly, predictions of the fabulous invalid's ultimate demise were finally beginning to seem less, well, dramatic than they had previously.

But Broadway made a big fat comeback. Prior to the 2008 global financial collapse, annual box office records for Broadway and the road were being broken year after year and had been on that path for over a decade with few exceptions. Many are calling it the dawning of a second Golden Age: houses are lit, people are working, and new, important and entertaining work such as *Spring Awakening* (2006), *Grey Gardens* (2006) and *In the Heights* (2008) continues to get produced, win Tony awards and lead the form in new directions.

In some ways, Broadway fans all over the world have Lloyd Webber, Mackintosh and Disney to thank. It was the success of spectacular British megamusicals in the 1980s that resuscitated popular interest in the stage musical – Broadway's biggest-earning genre by far. And it was megamusicals' Fordist models of reproduction that allowed for their successes all over the world. Disney then introduced a more fully cybernetic model of theatrical production to Broadway, which eventually brought in other new players to the street, too. Although for a time there was widespread concern that both British and Disney megamusicals might be crowding out other fare on the Rialto, newer producers have been smaller outfits nearly as often as they have been major media transnationals. And nearly as often, these smaller producers have chosen to produce daring shows over cautious, critic-proof fare.

There are still reasons for concern, however. Stage musical fans in cities with fewer theatres and smaller audience bases don't usually enjoy the same degree of variety as New Yorkers or Londoners. Megamusicals continue to dominate these markets; even in mature theatre centres like Toronto, Chicago and Sydney, there are often more *Lion King*s on the boards than *Spring Awakening*s.

What is more, performers working inside megamusicals remain frustrated by the interpretive delimitations imposed on them by megamusical methods of reproduction. Workplace alienation, of course, is nothing new to most of us, and even actors and musicians are accustomed to encountering it on dull, repetitive, poorly done gigs. Certainly Hollywood actors and studio musicians are accustomed to the realities of industrial routinization. But historically, the same cannot quite be said of Broadway's professional cohort, or that of the West End or Toronto or elsewhere. The point here is not that stage actors never suffer hardships at work, but that when they do suffer hardships, they are not generally associated with alienation born from drudgery. The myths of showbusiness inculcate performers and audiences alike to believe that if you choose Broadway, or the West End, or just 'The Theatre' as the place to build a career, you might never get rich, but you can at least expect a certain amount of job satisfaction.

But actors are finding this kind of satisfaction harder and harder to find. If one uses older shows such as *Phantom* or *Les Miz* as a benchmark – melodrama, anodyne diatonicism, no jokes – newer shows like *Hairspray*, *Dirty Dancing* (2006) and the Queen musical *We Will Rock You* (2002) are barely recognizable as megamusicals. Yet one interviewee with an intimate knowledge of the rather modestly budgeted *Hairspray* recently complained that 'the show is tracked within an inch of its life' both on stage and backstage. This means that every action of every actor, stagehand and wardrobe assistant is laid down in 'tracks' marked in space by tape on the stage floor and, more recently, on the floor backstage as well. Tracks have long been a part of Broadway production, but their new, heightened thoroughness and ubiquity suggest that you don't have to be a giant media company or Cameron Mackintosh these days to impose strict 'quality controls' on shows with plans for rollouts – large or small. Another interview subject working on the Toronto production of *Dirty Dancing* reported that Eleanor Bergstein, who wrote and co-produced both the movie and the new stage musical, understands that her new audiences for the stage production expect 'an exact reproduction of the film' and behaves accordingly in rehearsal. Still another relayed a conversation he'd had recently with a colleague preparing a big part in the Toronto production of *We Will Rock You*. 'How're you enjoying rehearsals?' my informant enquired. 'They're great!' his colleague replied. 'I just do what I'm told and nobody gets hurt.' For over 20 years now, Broadway's big fat comeback has owed a great deal to the influence of megamusical methods of reproduction, which have had an indelibly negative impact upon industry conceptions of actors and acting. 'For actors', as one prominent Broadway director put it to me, 'it's a loser's game.'

Meanwhile, old-school fans of the stage musical form complain long and loudly about the number of adaptations of movie properties playing the street. Although there's new, original work out there, so the objection goes, for every daring producer risking an *In the Heights* there are two *Legally Blondes* being mounted by corporate suits. All these adaptations are crowding out authentically *Broadway* Broadway. Where and when are the new *Broadway* classics to be born?

There can be no denying that adaptations of film properties are now scattered everywhere on the street, and that this is partly because producers – big and small – have finally embraced new digital logics and, as a consequence, are choosing bankable titles more and more regularly in order to reduce risk, and that this strategy often has predictably disappointing results. But if we really are to come to grips with the challenges that the new digital episteme presents to live-theatrical production, we first need to acknowledge that there's nothing inherently wrong with turning a movie into a show. To resist such adaptations categorically is to retreat to a kind of formalism that serves no real purpose. If its takes on Bergman's *Smiles of a Summer Night* (*A Little Night Music*, 1973) *Romeo and Juliet* (*West Side Story*, 1957), and Anita Loos' novel *Gentlemen Prefer Blondes* (*Gentlemen Prefer Blondes*, 1949) provide any indication, Broadway has been in the adaptation business for some time already. Clearly, these adaptations can be as innovative and original as any made-for-Broadway show if they're done right. Even today, when a show is first produced, composers, lyricists and designers continue to come together with directors, musicians and actors to collaborate in a rare and sometimes magical creative process. And when that process *is* magical, no matter what the story's original provenance may be, genuine Broadway is born.

It is in the moment of Broadway *re*-production, however, that newer, cybernetic forces of simulacral replication kick into gear, that the once-strong sense of the stage musical as a unique subject-experience is lost to the extension of the theatrical franchise and to its endless generation of spectacular object-commodities. This is the moment when both Broadway fans and Broadway professionals run into trouble. The subsequent push of the spectacular franchise is, either tacitly or explicitly, based upon a *template* of an original production whose cast and creative team have already completed all the exciting collaborative work of turning written page into breathing performance. The *template* of this collaboration, however, is not itself collaborative: it is starkly, soberingly reified, then licensed internationally. Such is the megamusical's sobering global legacy, and if it is not entirely uniform it is certainly widespread.

On stage in megamusical reproductions that continue to play everywhere, and in reproductions of the many smaller mainstream shows that are now emulating megamusical reproduction methods in small or larger measure, actors aren't breathing performances into the text anymore, they are mimicking the breathing of others instead. In 2008, performers on many mainstream stage musicals continue to confess to a sense of bewilderment best expressed to me back in the 1990s by a principal in the Toronto production of *Phantom of the Opera*. He wondered what it would be like to meet with his 12 other global counterparts 'and sit down and say, "Why do you suppose we do all this: move this way and this way and ...?"' He paused. 'Maybe it would only be the very original one that would really know.'

In such a fashion does the new Recombinant Broadway play endless games of broken theatrical telephone with its audiences, producing mounting conditions of affective meaninglessness on stages all over the world. These new conditions of affective meaninglessness can occur inside original Broadway fare, or on a Broadway show that enjoyed popular success as a movie first. Whether or not the show is an adaptation does not matter; whether or not it is a reproduction often does.

If the global economy of live theatre has always been complex, it has become increasingly so since the inception of the megamusical phenomenon back in the 1980s, and different logics remain discernable on different productions, large and small. But as the really big players continue to move new product into new territories, what is remarkable about their progress, and the progress of those who wish to emulate their successes, is how much the Fordist industrial logic, born in the days of *Cats*, *Phantom* and *Les Miz* (or what we may think of as the first phase of the theatre's global-industrialization), remains in operation. It is important to resist painting too determinist a picture here. Producers do incorporate techniques tuned to a more recent period of putatively post-Fordist flexible specialization. First-run licences often allow or even encourage local productions to insert local jokes; minor ameliorations to staging are often incorporated into a show's globally circulated production template. And certainly stage musical reproductions mounted in languages other than English complicate the narrative presented here in important ways. But these techniques really complement, rather than define, new normative practices.

Yet there are other respects in which Fordism is clearly not the only dynamic at work inside the global theatrical economy today. Certainly new digital technologies of spectacle don't rely on any kind of outdated, one-size-fits-all production logic. And a certain basic industrial 'flexibility' will always be associated with Broadway production because it is, after all, based on contract labour – contract labour on risky ventures, moreover, that tend to fail more often than they succeed. But there is now a new and notable double movement operating at the heart of the new theatrical commodity, one which guarantees that the core dynamic of theatrical global-industrialization necessarily remains one of global standardization and global routinization. Inside this double movement, producers seek first to produce and to market the unique, one-off, 'live' experience of a night at the theatre to audiences for a show's originating production. Then, for subsequent reproductions, they seek to *lock down* the live in the service of fetishistic replication. The movement is intuitive and sensible, not paradoxical, if the real stage in question is a new global market for the theatre characterized by 'liquid assets' and their convergent possibilities.

Epilogue

In 1936, Walter Benjamin famously described the collapse of 'aura' with the arrival of photography and cinema as mass forms and the concomitant consolidation of 'the age of mechanical reproduction'. According to Benjamin, aura's collapse entailed a change in the way people valued art objects. Previously, such objects had been valued for their singularity, their unique histories within a given time and place. Aura, for Benjamin, was a quality of historical testimony, one directly disputed by art objects' new status as *reproducible* objects inside new systems of mass image reproduction. Benjamin both celebrated and lamented aura's collapse. Whichever is more apt, the fact that Benjamin used opposing examples from the stage to make his larger arguments about the cinema, and aura's inevitable collapse there, is quite telling. Proclaiming, for example, that 'there is indeed no greater contrast than that of the stage play to a work of art that is completely subject to or, like the film, founded in, mechanical reproduction' (1968, 229–30), Benjamin sought to mark the theatre as a place where the assembly line was still kept at bay, and where actors were still in possession of their aura – what we may parse here as their singular signifying power – as a consequence.

With the arrival of theatrical Fordism, the theatre came fully and finally to inhabit the age of mechanical reproduction. And in precisely the manner that Benjamin foresaw the collapse of aura advancing in the cinema, but entirely against his predictions for the theatre, we have witnessed its collapse on the stage as megamusical actors, largely relieved of their interpretive function, are transformed into moving props or, in the words of megamusical professionals, into 'machines' and 'cogs', void of their own testimony.[3]

With the coming of theatrical global-industrialization, the stage actor's aura has collapsed, and with it the possibility of 'one-off' performances by individuals who, as 'one-off' human subjects, have laboured upon themselves to noble communicative purpose. Jean Baudrillard's (1983) term for Benjamin's moment of mechanical reproduction is 'the industrial simulacrum' – a 'second order' of Western epistemic organization following on from a pre-industrial, first order in which original objects were valued over counterfeit ones. Within series of things created inside the second, Benjaminian episteme, however, 'objects become undefined simulacra of one another'. The relation between two identical objects 'is no longer that of an original to its counterfeit ... but equivalence, indifference' (97). Blissfully unaware of either Benjamin's or Baudrillard's contributions on this matter, a Toronto music director put the historical progression to me this way: 'In the megamusical, everyone's an understudy.'

But even as the age of mechanical reproduction has finally emerged on stages across the planet, in production offices offstage, the age of *cybernetic* reproduction – of Baudrillard's *third*-order simulacra – has taken hold. There, producers organize schedules, coordinate with growing numbers of assistant and resident directors, and monitor in multiple ways the desired uniform consistency between shows scattered across vast distances in order to lock down the live. And whether producers work to do this consciously or otherwise, they do so nonetheless by looking both backwards and forwards into industrial history. Backwards by overseeing the efficaciousness of quality control techniques deployed by their staff in the Fordist, industrial simulacrum of the rehearsal hall and the theatre; forwards in the hope that such efforts will adequately tame the inherent volatility of their new global theatrical commodities. These commodities, volatile because people still try actually to act, to interpret, to get inside them, must nonetheless be controlled if they are to fit into the increasingly flexible distribution plans of new global producers – plans based on the post-Fordist, liquid realities of digital convergence and synergy. Relegating them to the same ontological status as branded merchandize, producers thus work to transform stage actors into so much 'software' – software that now requires loading into the new 'hardware' of conventional theatre spaces busily integrating with the logics of cybercapitalism. Stage actors aren't a perfect fit for these newly cybernetic containers, but if they are pushed hard enough, it seems, they can be squeezed in, provided worker disaffection is a negative externality producers are prepared to live with.

The question remains whether audiences are forever prepared to do likewise.

Notes

1. Dates indicate the year of the original stage production's premiere.
2. 'The road' is Broadway terminology for the circuit of North American cities that regularly host touring productions of Broadway fare.
3. In addition to Burston (2000) and Russell (2007), see also Philip Auslander (1999), whose discussion of how, inside our 'mediatized' social conditions, 'aura, authenticity and cult value have been definitively routed, even in live performance' (51), provides an important counterpart to these discussions.

References

Auslander, P. 1999. *Liveness: Performance in a mediatized culture*. London: Routledge.

Bart, P. 1996. The mouse fights back. *Variety*, 30 September–6 October: 11, 56.

Baudrillard, J. 1983. *Simulations*. New York: Semiotext(e).

Benjamin, W. [1936] 1968. The work of art in the age of mechanical reproduction. In *Illuminations: Essays and reflections*, ed. Hannah Arendt, 217–52. New York: Shocken Books.

Burston, J. 1998. Theatre space as virtual place: Audio technology, the reconfigured singing body, and the megamusical. *Popular Music* 17, no. 2: 205–18.

———. 2000. Spectacle, synergy and megamusicals: The global-industrialisation of live-theatrical production'. In *Media organisations in society*, ed. James Curran, 69–81. London: Arnold.

Canby, D. 1995. Look who's talking: microphones. *New York Times*, 22 January, Section 2: 1, 4.

Cox, D., and G. Evans. 1998. B'way rules rewritten to heed 'Lion's' roar. *Variety*, 22 December 1997–4 January 1998: 1, 78.

Cox, G. 2008a. Harvey enters from wings – Exec's hit movies move center stage as tuners. *Variety*, 23–29 June: 29, 33.

———. 2008b. Pic lays its love on 'Mamma': Adaptation a super trouper for Rialto, road. *Variety*, 28 July–3 August: 36, 39.

Diderot, D. 1883. *The paradox of acting*. Trans. Walter Herries Pollock. Preface by Henry Irving. London: Chatto & Windus.

Gubernick, L. 1994. Hollywood angels. *Forbes* 153, no. 11: 126.

Lister, D. 1995. The cloning of Andrew Lloyd Webber. *Independent*, 11 February: 25.

Moran, A. 1998. *Copycat television*. Luton: University of Luton Press.

Robertson, C. 2007. 'Mermaid' approaches, so 'Beauty' will close. *New York Times*, 18 January: B1, 7.

Russell, S. 2007. The performance of discipline on Broadway. *Studies in Musical Theatre* 1, no. 1: 97–108.

Siegel, T. 2008. H'W'D's musical mania. *Variety*, 6–12 October: 1, 63.

Sunhee, H. 2008. Korean pics pump booming tuner biz. *Variety*, 21–27 July: 31, 36.

Walsh, D., and L. Platt. 2003. *Musical theatre and American culture*. Westport, CT: Praeger.

Wolf, M.J. 1999. *The entertainment economy: How mega-media forces are transforming our lives*. New York: Times Books.

Global sport: Where Wembley Way meets Bollywood Boulevard

David Rowe and Callum Gilmour

Introduction: Importations and adaptations

Professional sport, initially a Western cultural form, is now a feature of all contemporary societies. Organized international competition, industry expansionism, technological development, and popular cultural diffusion have globalized sport participation and spectatorship. As sport has circulated the globe, its practice and expression have both replicated its established power formation (still substantially Western dominated) and, to varying degrees, challenged and modified it. For example, in the rapidly changing cultural landscapes of Asia, the nascent middle classes are now engaged in hybridized global/local cultures of sport consumption that cannot simply be dismissed as the outcomes of Western media and cultural imperialism. These shifts are manifest in the growing popularity of sports entertainment cultures that reflect the rise of emergent media and leisure economies, combining global aspirational cosmopolitanism with an assertion of local cultural identities and histories.

These multiple modes of cultural adaptation are evident in the ways in which globally diffused popular cultural products like English Premier League (EPL), association football (soccer) in general, and 'synthetic' local forms of global sport such as Indian Premier League (IPL) cricket, can be apprehended in various Asian contexts. This article will examine the production and consumption of local football, the EPL and IPL in Asia, particularly the manner in which they are negotiated, interpreted and contextualized within Asian socio-cultural milieux, in seeking to understand the complex structural and experiential dynamics of contemporary global sport. The EPL is a Western sport product that is vigorously marketed to Asia and other territories from the outside, while the IPL has been created within an Asian context out of the success of the Board of Control for Cricket in India (BCCI) in exploiting the political economic power derived from the size of the Indian cricket market. In both cases, and also in relation to locally practised football in Asia, global sport culture – especially fandom – is shown to be constrained by historical inheritance, yet engaged in the process of making new, multifaceted cultural histories.

Of particular interest to this discussion is the manner in which spaces of sports consumption and rituals of fandom are marked by global cultural flows while at the same time being reconstituted within Asian socio-cultural contexts. Asian modes of sport fandom, then, display the EPL's highly merchandized and commodified constructions of elastic fan identities in urban spaces such as shopping malls, live sites and food markets, and also ways of identifying with local clubs and national teams within the stadium that are ritualized, coordinated spectacles involving uniformity of colour, flag and banner waving, and synchronized mass movement. In India, the IPL has adopted Western-based sports entertainment practices such as cheerleaders, match-day live entertainment, and stadium branding to sell a domestic Indian sports league with many international players to both local and global audiences. The IPL is the first major city-based, franchise-organized sports league in India, and as such is representative of the nation's elevated position within world cricket and the global 'media sports cultural complex' (Rowe 2004). The IPL's strange hybrid of the English village green, 'Bollywood', and the Super Bowl strikingly reveals, then, how sport, as a mediated 'live' cultural form in which the in-stadium audience is a key part of the spectacle, is subject to multiple and sometimes bizarre cultural adaptations.

Sport, as a form of popular culture, is shaped by constantly mutating interactions between the media and sport industries and the established and evolving fan cohorts on which they depend (Horne 2006). The formation of significant organized domestic football fan cultures in East Asia, the development of the new IPL cricket competition and the extensive pan-Asian television audiences aggregated by EPL football, are by-products of new lifestyle modes that have emerged in the wake of economic and cultural modernization across significant parts of the Asian continent since the mid-twentieth century. Japan's post-Second World War economic transformation, and the subsequent emergence of its highly consumptive middle class, heralded the onset of a paradigmatic shift based on 'developmentalism' that, with its compelling progressivist-modernist narrative, would substantially redraw the socio-economic landscape of much of Asia. The economies and cultural consumption patterns of South Korea, Singapore, Hong Kong and Taiwan industrialized and liberalized from the 1960s onwards, followed in uneven ways by Thailand, Malaysia, Indonesia and the Philippines in the 1970s. Asia's most populous nations, China and India, rapidly emerged in the 1980s and 1990s as key sites of production, which in turn created the conditions for heightened domestic consumption of goods and services (Yamazawa 2002, 2), not least those relating to sport and leisure.

There is a tenable proposition that 'consumption behaviour creates the primary cultural context in which the new middle classes in Asia operate and provides the linkage between globalization and local urban cultures' (Clammer 2003, 403). This position informs the following discussion of sports cultures in Asia, which is framed here in terms of a global–local dialectic and corresponding hybridized and indigenized (Appadurai 1996) modes of consumption. Sport, we argue, is especially well suited to such cultural adaptations and reconfigurations, given the protean, mobile nature of its organizations, texts and practices.

Sport and glocalization in emerging Asian leisure economies

The development of organized professional sport within Asian nations (Rowe and Gilmour 2006) and the proliferation of Western sports across an array of domestic and pan-Asian media platforms attest to a burgeoning regional leisure economy. The demand for leisure and entertainment products has grown concurrently with the ascendancy of consumption capitalism and the growth of the middle classes within Asia, enabling the exhibition of social status, the performance of cosmopolitanism (Lozada 2006) and a conspicuous engagement with modernity. The potent, yet uneven and at times contradictory mix of a highly pervasive mass media, image-based marketing and consumerist lifestyle modes, operates within the many cultural formations of Asia that have produced the conditions for sporting glocalization. This now familiar concept – the product of Japanese marketing strategies – is described by Giulianotti and Robertson (2007) as:

> critically transcend[ing] the banal binary oppositions associated with globalisation, and so register[ing] the societal *co-presence* of sameness and difference, and the intensified *interpenetration* of the local and global, the universal and the particular, and homogeneity and heterogeneity. (168)

In examining three Asian spaces of sports production and consumption – (1) the pan-Asian reception of English association football; (2) football fan cultures in Japan; and (3) an Indian cricket league – we seek to reassess Appadurai's (1990) claim that 'at least as rapidly as forces from various metropolises are brought into new societies they tend to become indigenized in one or other way' (295).

The English Premier League in Asia and the 'self-inventing trans-national fan'

Traditional football fan culture is founded on an almost obsessively topophilic identification with 'place' that has remained acutely relevant at the sport's geo-economic 'centre' in Europe, despite football's intensive commercialization and global expansion to peripheral and emerging markets. For over a century, public engagement with professional football in the United Kingdom (including, but not confined to, England) has been characterized by a heightened and often performative identification with locality that is manifest in shared rituals and practices. The generational transference of collective memories and reverence for teams' histories and traditions – most of which only stem directly from the late Victorian era, but often with elements that can be traced to antiquity – created a strong sense of territoriality (Elias and Dunning 1986; Mason 1989), sometimes even of a violent kind (Taylor 1971). The increasing globalization of European professional football in recent years has not diminished this topophilia among English football fans, 'with clubs persisting as meaningful sites for ritualized identification with a geographical locality' (Horne and Manzenreiter 2008, 359). In part, this phenomenon is a reactionary and parochial response to the increasing internationalization of the professional game in England (Ruddock, Hutchins, and Rowe 2008). It is a significant by-product of a rapid increase in foreign capital investment in English clubs, a surge in foreign labour in the form of players, managers and coaches, and, of most significance here, the media-enabled aggregation of substantial global fan bases, particularly in the Asian region.

As these transformative influences have been visited upon top-flight English football in an unprecedented manner, the restless expansionism of the EPL and its major clubs has led to the aggressive pursuit of fan support that exists far beyond the geographic bounds of their immediate regions. The trans-national scope and proliferation of media technologies, combined with increasingly pervasive merchandizing activities, have fostered fan cultures and viewerships that eschew the norms, rituals and 'rules' that still inform much domestic English football fandom. In exchange, fan engagement 'from afar' is more concerned with the pursuit and display of status and prestige within the emergent consumer cultures of urban Asia. As Binnie *et al.* (2006, 14) state:

> The new middle and gentrifying classes are definable through their particular combination of economic and cultural capital that enables them to distinguish themselves from other classes on a number of fronts. In particular, it is the production and deployment of cultural capital and aesthetic sensibilities that provides the mechanism through which this class fraction finds its identity.

The manner in which the media-enabled EPL fan base in Asia connects with English football is through image-based marketing and commodity culture within wider narratives of economic development and aspirational cosmopolitanism. In England, proximity to the 'sacred site' of the home stadium is a key marker of fan authenticity within a hierarchical discourse of fans competing for legitimacy and standing within a wider supporter 'community'. Clear practices and circumstances exist within this value framework that limit a fan's claims to authenticity, in particular the fashion-based consumption of team-branded merchandizing and a large geographical distance between the fan and the team's principal location. Indeed, the negative association with 'distance' is heightened within England's 'big four' team fan communities: Manchester United, Liverpool, Arsenal and Chelsea. These successful English clubs command significant fan bases well beyond the bounds of their home cities and regions. So, for example, 'Cockney Reds' (London-based Manchester United supporters) are often disparaged by 'roots' supporters in team 'fanzines' for their lack of genuine spatial and social attachment to the northern industrial milieu.

The Asian EPL audience is primarily a media construction. This predominantly visually consumptive culture is only loosely connected to the partisan rituals informing European (and notably English) football fandom. Asian fan groups are frequently denigrated by England-based supporters (even more than those within the United Kingdom who do not come from or inhabit the originary space of the team) and so are even more disconnected and actively blocked from the channels of organized fan communication. When contact does occur (usually via electronic media such as the Internet), it is often combative and ultimately dismissive of the position of the distant, mediated fan. Such prejudice amongst many domestic English supporters is manifest in the deployment of the derogatory term 't-shirt' to describe Asian players believed to be signed by English clubs solely for their ability to drive merchandizing sales in Asia (Australian Broadcasting Corporation 2007). Additionally, the perceived 'elasticity' (Rowe and Gilmour 2008, 2009) of Asian fan loyalties further elicits disdain amongst some English-based fans, as is evidenced here in a posting on a Manchester United Bulletin Board from May 2008:

> I once saw a Korean with an Arsenal backpack, a United shirt and a Chelsea hat on. What's more, he was taking pictures of Sir Matt's [Busby, the revered former team manager] statue outside Old Trafford. That really does take the piss. (KingEric7 in RedCafe 2008)

This quotation makes reference to the widely held perception that Asian football fans have questionable, or at least 'flexible' loyalties and that the bond between them and the club is largely superficial and driven as much by fashion and brand marketing as by genuine fan fervour. In the same discussion thread, entitled 'Do most of our fans actually acknowledge our real rivalries?', an argument ensues between supporters who claim to be regular attendees of Manchester United home matches and a selection of overseas-based fans pilloried for their putatively lax loyalties, negligible cultural capital, and lack of historical engagement with the club and its traditions. Typical of such claimed superior positionality is the following remark:

> The first thing you should do as a foreigner when you CHOOSE a club is to learn about the history. You should take the word of the oldies and the match-going fans. I did from my friend's dad and I still give priority to words of caftards [subscribers to the RedCafe Bulletin Board] like topper, Fred and Ralphie. (Dippersripper in RedCafe 2008)

In emphasizing the word 'choose', this respondent asserts an intrinsic difference between 'local' Manchester United fans who are compulsorily so from birth, and international supporters who have selected their team on the basis of brand value, aesthetics, image and celebrity appeal. The distance – historical, geographical, social and cultural – between contexts diminishes the intense tribalism of English football. As a result, strict segregation of fans from rival teams (usually heavily policed by state authorities) is also reduced, producing instances of intermingling in the social spaces where Asian (and other international) EPL fan groups congregate.

The following Bulletin Board excerpts from international fan posts illustrate the glocalized nature of their engagement with football teams and other fans, and the differing priorities and social circumstances that inform their 'self-inventing trans-national fandoms' (Giulianotti and Robertson 2007, 177) that necessarily adapt to the very different circumstances in which their fandom is practised:

> I mean, come on, you're demanding that people across the globe from you adopt the same, trivial, sectarian feelings as you. (Klogon in RedCafe 2008)

> [A]s you've defined 'fan', you'd have to know a bit about the club's history. That's fine, as you could argue that that group is qualitatively different from the non-history-learning fans. But we'd need a new term for those non-fans who buy the merchandise, follow the results, are emotionally invested in the club's success, etc. Because merely branding them as 'non-fans' would group them with those who haven't even heard of United, and this grouping would be misleading. (Mr Wood in RedCafe 2008)

Such contestation over football fandom status and its associated hierarchies and meanings signals both the pressure for adaptation, and resistance to it, on the shifting terrain of the global media sports cultural complex.

EPL consumption and aspirational cosmopolitanism in the modern Asian city

The Asian EPL fan base discussed above exists within a peculiarly post-modern socio-cultural space in which global or Western-inspired capitalist modernity is indigenized within Asian contexts – with the global–local dialectic reflected in hyper-commercialized, deeply consumerist and increasingly affluent and urbanized lifestyle modes (Ang and Stratton 1996; Kahn and Kellner n.d.). Within a space undergoing transformation through accelerated economic growth (somewhat stalled in 2008 and for the immediate future) and rapid socio-cultural transition, the highly mediated and merchandized cultural form of the EPL provides an accessible outlet for the practice of a decidedly visual form of engagement with 'thick cosmopolitanism':

> Media consumers, simultaneously imbibing print, electronic, and satellite communications around the globe, come to imagine themselves as cosmopolitan participants in global commodity culture – surmounting the spatial constraint of locality, entering the global scene by means that deny geographic immobilities. (Schein 1999, 345)

In societies emerging from sustained periods of relative economic and cultural isolation, therefore, the EPL offers ready access to a high-profile element of global popular culture.

As Manzenreiter (2006) notes, 'football replica shirts can be read as nodes that connect the social experience of living in late modernity with global capitalism' (157). Thus Asian consumption of English football represents a vivid expression of collective and individual possession of capital. Despite such strategic uses of international cultural capital through conspicuous consumption and command over knowledge of non-indigenous sport, it is important to recognize the complexity of such 'self-inventing trans-national fandoms', particularly as the 'distinction between "traditional" and "consumer" fans fails to recognise the diverse and multiple ways individuals can connect with their chosen interest' (Crawford 2004, 31).

The intensive economic development of recent decades has radically altered the physical complexion of urban environments within Asia, which are now characterized by the ubiquitous presence of commodity culture iconography. As Clammer (2003) notes, 'it is the cultures of urban spaces that are most immediately and directly influenced by globalisation (in terms of consumption patterns and tastes, fashion, architecture, media and new forms of material culture)' (403). Already media- and advertising-saturated public spaces within Asia are heavily marked by the signs of the global media sport industry. Sportswear brands such as Nike and Adidas, and sport celebrity endorsers like David Beckham and Tiger Woods, are highly visible in Asian streetscapes. The EPL's determined advance into national and regional sports markets across the continent, then, is seeking both omnipresence and stylistic appeal.

International football consumption in Asia, as in more 'mature' sport zones, is centred primarily on social rituals now permeated by television. The majority of matches are broadcast live in prime time on Saturday nights, taking advantage of convenient time zones and further insinuating sport into the rhythms of mediated leisure. Televised EPL is, as a result, built into domestic spaces and also those that revolve around Asian night-time economies in bars, coffee shops, night-markets, restaurants, shopping centres and various live sites. A highly collective mediated experience, particularly in Southeast Asia, the EPL is widely viewed at local outdoor food and beverage outlets that form significant social hubs within countries like Singapore and Malaysia. The matches themselves, however, are watched in vastly different ways – sometimes occupying the centre of partisan interest, and at other times operating as merely background 'wallpaper' for other social activities. The sociality of televised sport viewing (Rowe and Stevenson 2006) is therefore as evident in Asia as in other global contexts, and as influenced by the specific conditions under which it is experienced.

Performance and mimicry: Football fan cultures in Japan

The most developed local football industries and fan cultures exist within the continent's most established economies in East Asia – notably Japan and Korea, and to a lesser extent China. Korea's K-League was established in 1983, Japan's J-League in 1993, and China's Jia League (now Chinese Super League) in 1994. To varying degrees, these leagues have successfully integrated themselves into their respective leisure economies. With Korea and Japan's co-hosting of the FIFA World Cup in 2002 and the progression of their national teams into the knockout stages of the tournament (alongside China's inaugural participation), East Asia has established itself as the continent's most successful region on the field of play. Of more interest here, though, is the manner in which Japan's professional football (the J-League) has commanded a significant female audience and cultivated a distinctive hybrid fan culture that has adapted the highly performative and spectacularized European 'ultra' supporter style to Asian socio-cultural contexts.

Despite the proliferation of the EPL on Asian television screens, it is in fact the more visually spectacular approach of the Continental European 'ultra' tradition that has had the most significant influence on Japanese professional football fan-style cultures. This highly organized and participatory style of team support (itself derived and adapted from English fan culture) involves rival fan groups competing to produce the most spectacular in-stadium display of team devotion, positioning spectators as active participants in the wider production of the media sport spectacle. The ultra tradition has been glocalized within Japanese football by means of its incorporation with the indigenous supporter culture known as *oenden* (Campion 2005). Common across multiple forms of Japanese popular culture, *oenden* is a highly visual, coordinated and vociferous mode of audience participation that is as common at pop concerts as at baseball matches. The *oenden* fan tradition has been integrated into the supporter cultures of the J-League, where it is manifest in highly expressive, ritualized and synchronized fan performances involving the grandiose display of team colours, flags and banners. As a result, 'Colourful clouds of smoke, posters of Che Guevara, samba rhythms and Japanized versions of "You'll never walk alone"' (Manzenreiter 2008) permeate the football stadia of Japan. These performative simulacra, almost entirely detached from their original cultural reference points, involve a curious pastiche of European supporter styles. They entail a collective display of coordinated movement with an almost totalitarian aesthetic, as well as an expression of emerging individualism in often restrictive, historically communitarian Asian societies (Cho Han 2004).

This heavily branded football culture is designed to appeal both to the youthful 'fandom generation' (Cho Han 2004, 11) that permeates East Asian popular culture and to the sport's increasingly feminized supporter base (Kim 2004; Tanaka 2004; Manzenreiter 2008). While Japan's previously booming economy experienced a significant contraction in the 'lost decade' of the 1990s, its creative economy enjoyed an unprecedented period of global expansion (Nakamura 2003). A significant feature of this creative boom was the promotion of an 'idol' culture (Campion 2005) that fetishized celebrity. Incorporated into the consumption of football in East Asia, this development prompted one commentator to remark that 'today's soccer fans are like fans watching their idols – soccer fandom needs a new definition' (Lai 2003, D9). This form of 'idolatry' is represented within Japanese football through promotion and image-based marketing of the sport's biggest and most marketable stars, and their endorsed goods, to an enthusiastic female fan base:

> Players are evaluated by whether they are 'good looking' (*iketeru*) or 'bad' (*iketanai*), or can be 'prince-like' or 'pretty' (*kawaii*). Some of the aforementioned modes of watching soccer are cross-cultural in the sense that they commonly share some elements with other fan cultures. They may resemble the ways in which boys' pop groups and visual bands are chased and loved by their followers. Buying the dedicated magazines and photo-books and watching the video tape that records their television experience. (Tanaka 2004, 55-6)

This combination of merchandizing through a set of highly developed team identities, colour schemes and logos, combined with the intensive 'celebritization' of players' personalities, has specifically been designed to mobilize the consumptive power of Japan's young female market. A highly gendered experience has been created around Japanese football whereby female desire is commodified by promoting fantasy and romance around star players (Kim 2004), and by explicitly coding previously grim, male-dominated stadia as bright, woman-friendly spaces of 'pleasure and fun' (Cho Han 2004):

> Female supporters express their own distinctive fashion and support. Sun-tanned high-school football girls occupy one corner of the stand, whereas housewives enjoy their rare free time on the other corner. Some high-school girls wear the team shirt over their school uniform.

This 'fashion' can be observed every Saturday afternoon at a J-League match. During half-time their mobile phones work at full throttle. They exchange what they saw, thought and experienced about the game and players with friends and family in other stadiums and at home. (Tanaka 2004, 54)

Football fan culture in East Asia exists as a significant site of gendered cultural articulation and appropriation, whereby the traditionally masculine-coded spaces of professional football (stadia, television coverage, sports magazines, and so on) have been reconfigured to meet the demands of a significant female audience. As Manzenreiter (2004) notes in relation to television coverage of the game, 'the Japanese media's obsession with star players', a by-product of the game's feminization, is reflected in the enormous 'emphasis on close-up shots and zoom-ins on popular players' (302). This televisual development is illustrative of the significance of 'female pleasure' as of critical exchange value within the Japanese football industry, where nearly 45% of all match attendees are female (Manzenreiter 2008, 248). The commodification and objectification of the male body, 'inverting the traditional relationship between the female body and the male gaze' (248), has resulted from the positioning of heterosexual women as central to football fandom, thereby providing an unusually mainstream medium for the expression of traditionally repressed female sexuality in East Asian society. The third 'case study' in this article addresses a related form of sport and celebrity entertainment that represents a very recent but influential development in the media sports cultural complex – the sporting innovation that is Twenty20 cricket.

The Indian Premier League, Bollywood and the incorporation of the Western sports entertainment model

The Indian Premier League (IPL) (like Indian cricket in general) offers perhaps the most tangible example yet of a genuine political and economic shift in the power relations of a major sport from West to East. The rise of the Indian economy over the past two decades, combined with the nation's population in excess of one billion and its elevation of cricket above all other sports (including hockey), has placed India at the centre of the sport (Gupta 2004; Rumford 2007). India's geo-political economic dominance of cricket has prompted Majumdar to conclude that it is 'close to a truly postcolonial sport' (2007, 89).

The genesis of Indian cricket's financial pre-eminence lies with the deregulation of the Indian media market, which had once been dominated by a sole television source – the state-run Doordarshan network. The deregulation and subsequent market entry of multiple national, regional and global broadcasters such as Rupert Murdoch's ESPN Star Sports, Zee Sports and Sony TV paved the way for what is now an extraordinarily competitive contemporary television landscape (fuelled by a vast legal and illegal betting market) in which the television rights for blue ribbon content – such as Indian cricket – are sold at a very high premium. As recently as 1993, the BCCI sold the television rights for a series against England for US$40,000 (Premachandran 2008a), but by 2008 it received US$1.1 billion in television income for its IPL property (Tabakoff 2008), and another US$1 billion for the related Twenty20 Champions League tournament (Sweney 2008).

The financial ascendancy of Indian cricket has allowed the BCCI to exert enormous political influence over cricket's national, regional and international administrative bodies:

In the case of international cricket the nonwestern nations have established greater control over the game and look like determining the future course that the game will take. They are likely to decide its format, its content, (and) its venues. The organisation, economics, and character of the game is now dominated, and being changed by the nonwestern world. In that sense, the game is somewhat unique to the globalization process because it reflects changes in the power structure of international organization that typically do not occur in the sphere of trans-national organized activity. (Gupta 2004, 273–4)

The most significant manifestation of India's ascendancy is the formation of a new cricket competition (IPL) that aspires to be the cricketing version of EPL football – that is, a domestic league with a global television audience performed by the world's highest profile players. Launched in early 2008 with some success, the IPL dramatically highlights the emergence within India of a sizeable consumptive middle class. As Andrew Wildblood (a senior employee of global sports agency IMG) suggests, the IPL caters to an expanding leisure and entertainment economy: 'India is a rapidly growing economy with an emerging middle class, but if you want to go out here you go to a movie and that's about it. There's a huge demand for entertainment, and we are providing the perfect product model' (qtd in Kelso 2008).

The IPL has appropriated much of the marketing and promotional expertise, branding strategy, merchandizing and media packaging that now define modern Western sports production – and transplanted them to particularly Indian socio-cultural spaces. Significantly, however, the capital generated by the IPL is distributed within India rather than expropriated from it, and Indian cricket's economic power (approximately 80% of total global turnover in the game) has in many respects positioned the BCCI as cricket's unofficial ruling body, and so the principal shaper of the sport's contemporary image. As one observer has put it: 'The Indian board has owned the game for years, but now, thanks to the stratospheric popularity of Twenty20 cricket, it has found the means to shape it in its own image' (Miller 2008). The International Cricket Council (ICC), the international governing body, has increasingly come under the control of India, assisted by its (sometime) strategic subcontinental allies, Pakistan, Sri Lanka and Bangladesh (Williams 2003). This governmental hegemony, derived from economic power, has enabled India to overcome Western – especially English – domination, as is clearly signified by the relocation of the ICC in 2005 from Lords in the capital of England to Dubai in the United Arab Emirates.

While the IPL is reflective of contemporary Western sports frameworks, it incorporates the highly stylized visual aesthetic and potent celebrity cachet of Indian cinema culture (popularly known as 'Bollywood') within its match-day entertainment and team branding. This 'Bollywoodization' of cricket, involving the merging of Indian sport with the entertainment values of its film industry, has diversified the traditional cricket audience, drawing more female and family-oriented spectators and diluting the traditional partisanship associated with Indian cricket crowds. As one observer noted:

> I was bemused to see almost every replica shirt embossed with Khan-12. Khan, for those that live on Mars, is Shah Rukh Khan, India's most popular actor and the owner of the franchise. The Shah Rukh shirt, which I've spotted in other Indian cities as well, says a lot about the new breed of fan that's watching the IPL games. Kolkata's sports fans are supposed to be the most passionate in the country, yet the manner in which they've ditched Ganguly for Shah Rukh suggests that the hardcore sports-lover is staying away. (Premachandran 2008b)

The IPL schedule is organized along the lines of the US sports entertainment model, with matches played on a nightly basis throughout the duration of the season in order to provide maximum prime-time content for its television partners. Teams are based in major urban centres such as New Delhi and Mumbai rather than in the more traditional (for India) regions or major companies. In addition, the IPL is structured according to the US-based sports franchise system, with team names such as Daredevils and Kings, cheerleaders imported from the United States (also used to train local women, who adapted cheerleading styles to reflect local cultural norms and presentational regimes) and franchises awarded to the highest bidders (in a number of cases, Bollywood actors) in a highly publicized (and lucrative) team auction run by the BCCI. In terms of playing talent, the IPL reverses the conventional West–East flow of the new international division of cultural labour (Miller et al. 2003) that has seen Asian stars from other sports such as football migrate to wealthier Western leagues in order to earn superior wages. Instead, the IPL has contracted the world's leading players with salaries unprecedented in the history of cricket.

Perhaps, though, the most striking innovation involving the IPL is the structure of the game itself. The code of cricket used in the IPL is Twenty20, a recent English invention shortening the austere format of long-form Test match cricket, and even more abbreviated than the one-day game invented (also in England) in the 1960s as a more convenient and lively spectatorial experience. While Test matches run for five days and often take place during business hours, Twenty20 cricket lasts for approximately three hours after the conventional working day has ended, and is thus constructed to fit more easily into leisure-time patterns and prime-time television schedules. This 'truncated' format's televisual nature is also evident in the spectacularized version of cricket that its rules facilitate – valorizing strength and power hitting rather than the more subtle variations and skills that characterize Test cricket, and focusing strongly on celebrity cricketers and team owners. IPL matches are similar in length to Bollywood films – and comparably colourful and star-driven vehicles.

Conclusion: Sporting hybrids and indigenizations

In this article, we have selected and examined some significant elements of the dynamic media sports cultural complex. We have sought to demonstrate that the trajectory of sport's global cultural development is not subject to a simple logic of adoption and imitation, but is created out of multiple, intersecting correspondences – and non-correspondences – between histories, sites and social formations. A range of processes can be offered up as shaping the cultures of global sport, including globalization itself, Americanization, televisualization and others (Miller et al. 2001). In concentrating on differential contexts, though, we have emphasized the ways in which such mega-processes operate on, and interact with, sporting phenomena in myriad culturally adaptive ways.

We have sought to demonstrate that the EPL is successfully exported to Asia, and that its consumption there is derided by many traditional fans from its place of origin, but is nonetheless refashioning the practice and meaning of football fandom. Football fan cultures in Japan were shown to draw on local fan traditions grafted on to style-based consumption and strategic feminization, while the IPL registered a pronounced shift towards India as the dominant power in cricket and marked out Bollywood as its most adaptable and globally recognized popular cultural form. Much has been made of the 'decline of the West' and the rise of the Asia-Pacific region in the fields of politics and trade. It is apparent that, accompanying such global realignments, the cultural complexion of sport – not least in the vibrantly performative sphere of fandom – will undergo profound change. Established analytical frameworks foregrounding sporting cultural importation and pale imitation require a concomitant reconsideration and adaptation.

References

Ang, I., and J. Stratton. 1996. Asianizing Australia: Notes toward a critical trans-nationalism in cultural studies. *Cultural Studies* 10, no. 1: 16–36.

Appadurai, A. 1990. Disjuncture and difference in the global cultural economy. *Theory, Culture and Society* 7, nos. 2–3: 295–310.

———. 1996. *Modernity at large: Cultural dimensions of globalization*. Minneapolis: University of Minnesota Press.

Australian Broadcasting Corporation. 2007. English football. *Saturday Extra*. http://www.abc.net. au/rn/saturdayextra/stories/2007/2014719.htm

Binnie, J., J. Holloway, S. Millington, and C. Young. 2006. Introduction: Grounding cosmopolitan urbanism: Approaches, practices and policies. In *Cosmopolitan urbanism*, ed. J. Binnie, J. Holloway, S. Millington, and C. Young, 1–34. New York: Routledge.

Campion, C. 2005. J-Pop's dream factory. *The Observer*, 21 August. http://www.guardian.co.uk/music/2005/aug/21/popandrock3

Cho Han, H. 2004. Beyond the FIFA World Cup: An ethnography of the 'local' in South Korea around the 2002 World Cup. *Inter-Asia Cultural Studies* 5, no. 1: 8–26.

Clammer, J. 2003. Globalization, class, consumption and civil society in Southeast Asian cities. *Urban Studies* 40, no. 2: 403–19.

Crawford, G. 2004. *Consuming sport: Fans, sport and culture*. London: Routledge.

Elias, N., and E. Dunning. 1986. *Quest for excitement: Sport and leisure in the civilizing process*. Oxford: Basil Blackwell.

Giulianotti, R., and R. Robertson. 2007. Recovering the social: Globalization, football and trans-nationalism. *Global Networks* 7, no. 2. 166–86.

Gupta, A. 2004. The globalization of cricket: The rise of the non-West. *International Journal of the History of Sport* 21, no. 2: 257–76.

Horne, J. 2006. *Sport in consumer culture*. Basingstoke: Palgrave Macmillan.

Horne, J., and W. Manzenreiter. 2008. Football, *komyuniti* and the Japanese ideological soccer apparatus. *Soccer and Society* 9, no. 3: 359–76.

Kahn, R.V., and D. Kellner. n.d. Global youth culture. http://www.gseis.ucla.edu/faculty/kellner/essays/globyouthcult.pdf

Kelso, P. 2008. Game's future at stake as world's top players go to the highest bidder. *The Guardian*, 20 February. http://www.guardian.co.uk/sport/blog/2008/feb/20/gamesfutureatstakeasworld

Kim, H.M. 2004. Feminization of the World Cup and women's fandom. *Inter-Asia Cultural Studies* 5, no. 1: 42–51.

Lai, P.F. 2003. Our soccer field. *Ming Pan*, 10 August.

Lozada, E.P. Jr. 2006. Cosmopolitanism and nationalism in Shanghai sports. *City and Society* 18, no. 2: 207–31.

Majumdar, B. 2007. Nationalist romance to postcolonial sport: Cricket in 2006 India. *Sport in Society* 10, no. 1: 88–100.

Manzenreiter, W. 2004. Japanese football and world sports: Raising the global game in a local setting. *Japan Forum* 16, no. 2: 289–313.

———. 2006. Sport spectacles, uniformities and the search for identity in late modern Japan. *Sociological Review* 54, no. 2: 144–59.

———. 2008. Football in the reconstruction of the gender order in Japan. *Soccer and Society* 9, no. 2: 244–58.

Mason, T. 1989. *Sport in Britain: A social history*. Cambridge: Cambridge University Press.

Miller, A. 2008. Opening a Pandora's box. *CricInfo*, 2 February. http://content-www.cricinfo.com/magazine/content/current/story/334467.html

Miller, T., G. Lawrence, J. McKay, and D. Rowe. 2001. *Globalization and sport: Playing the world*. London: Sage.

Miller, T., D. Rowe, J. McKay, and G. Lawrence. 2003. The over production of US sport and the new international division of cultural labour. *International Review for the Sociology of Sport* 38, no. 4: 427–40.

Nakamura, I. 2003. Japanese pop industry. Stanford University Japan Centre. http://www.stanford-jc.or.jp/research/publication/DP/pdf/DP2003_002_E.pdf

Premachandran, D. 2008a. The ire of the tiger. *Sydney Morning Herald*, 2 February. http://www.smh.com.au/news/cricket/the-ire-of-the-tiger/2008/02/01/1201801033663.html

———. 2008b. The long and the shirt of replica sports clothing. *The Guardian*, 28 May. http://blogs.guardian.co.uk/sport/2008/05/28/the_long_and_the_shirt_of_repl.html

RedCafe. 2008. Do most of our fans actually acknowledge our real rivalries? [Discussion thread]. *RedCafe2.net* [online forum], 19 May. http://www.redcafe.net/f6/do-most-our-fans-actually-acknowledge-our-real-rivalries-203304

Rowe, D. 2004. *Sport, culture and the media: The unruly trinity*. 2nd ed. Maidenhead: Open University Press.

Rowe, D., and C. Gilmour. 2006. The future role of professional sport and the media in the Asia Pacific societies. In *Asia Pacific yearbook* [Anuario Asia-Pacífico 2005, Edición 2006], 473–81. Barcelona and Madrid: Elcano Royal Institute, Casa Asia and the CIDOB Foundation.

———. 2008. Contemporary media sport: De- or re-Westernization? *International Journal of Sport Communication* 1, no. 2: 177–94.

———. 2009. Sport, media and consumption in Asia: A merchandized milieu. *American Behavioral Scientist* (forthcoming).

Rowe, D., and D. Stevenson. 2006. Sydney 2000: Sociality and spatiality in global media events. In *National identity and global sports events: Culture, politics, and spectacle in the Olympics and the football World Cup*, ed. A. Tomlinson and C. Young, 197–214. New York: State University of New York Press.

Ruddock, A., B. Hutchins, and D. Rowe. 2008. Respatialising sport fandom: Digital mediation, connectivity and MyFootballClub. Unpublished paper.

Rumford, C. 2007. More than a game: Globalization and the post-Westernization of world cricket. *Global Networks* 7, no. 2: 202–14.

Schein, L. 1999. Of cargo and satellites: Imagined cosmopolitanism. *Postcolonial Studies* 2: 345–75.

Sweney, M. 2008. Twenty 20 Champions League rights snapped up by ESPN in $1bn deal. *The Guardian*, 11 September. http://www.guardian.co.uk/media/2008/sep/11/sportsrights.television

Tabakoff, N. 2008. Loyalties face test over T20 millions. *The Australian*, 16 January. http://www.theaustralian.news.com.au/story/0,25197,23059503-5001505,00.html

Tanaka, T. 2004. The positioning and practices of the 'feminized fan' in Japanese soccer culture through the experience of the FIFA World Cup Korea/Japan 2002. *Inter-Asia Cultural Studies* 5, no. 1: 52–62.

Taylor, I. 1971. Soccer consciousness and soccer hooliganism. In *Images of deviance*, ed. S. Cohen, 134–64. Harmondsworth: Penguin.

Williams, J. 2003. 'Paki cheats!' Postcolonial tensions in England–Pakistan cricket. In *Sport and postcolonialism*, ed. J. Bale and M. Cronin, 91–105. Oxford: Berg.

Yamazawa, I. 2002. Economic development and structural changes in East Asia: Overview. Asia-Latin America cooperation seminar promoting growth and welfare: The role of institutions and structural changes in Asia, 29–30 April, in Santiago, and 2–3 May, in Rio de Janeiro. http://www.eclac.org/brasil/noticias/noticias/4/9794/yamazawa2904.pdf

Commercialization and culture in Australian gambling

Richard Woolley

As the legend goes, Australians will gamble on anything. As Peter Charlton (1987) captured in the title of his book on the subject, even 'two flies up a wall'. According to this way of thinking, there is something of an 'elective affinity' between the Australian character and gambling. Perhaps this is related to the conditions encountered by the British colonists, with the post-1788 history of Australia being replete with stories of hardship and danger, stories that seemingly turn on nothing more than 'blind luck'. An interpretation of the Australian passion for gambling as being formed via a particular experience of contingency, based in the pragmatics of struggle, situates the culture of gambling in Australia in a specific socio-historical context. This contextualization contrasts with a relatively intellectual or spiritual apprehension of the role of number and luck in deep structures of fortune and fate, as found in Chinese culture for example (Basu 1991).

If there is something in the legend of a particularly 'Australian' attachment to gambling, it is refracted through different social and cultural prisms (McMillen 1996). As John O'Hara (1988) has argued, the stories of early Australian gambling cultures are tied up with the construction of working-class experience in Australia's young cities. This experience was not always legitimate, or viewed as desirable. The 'right' to gamble was won in the colonies despite strong religious opposition to a practice construed as a vice or moral affliction (O'Hara 1988). The importance of working-class male experience to the flourishing of gambling continued to be important after the Second World War, particularly as the poker machine industry emerged in New South Wales (Lynch 1990). However, after 1956, when the NSW club sector began to legitimately operate poker machines, gambling also became available for women, and slowly grew more socially acceptable.

The period of Australia's gambling history of interest here is the phase of liberalization and expansion of commercial gambling, which accelerated from the 1970s. There were two main aspects of this expansion. State and territory governments licensed and continue to control casinos. In the 1990s, poker machines were introduced to pubs and/or clubs in all parts of the country except Western Australia. This proved to be somewhat fortuitous timing – the emergence of commercialized gambling as a multi-billion dollar industry coincided with almost two decades of uninterrupted economic growth and rising personal and household wealth. Commercialized gambling was thus able to situate itself as a form of leisure and entertainment available to citizens of an affluent society. However, at the same time this society also became more complex, with a diverse multicultural nation emerging. The intersection of differing cultural values and practices with the flourishing opportunities to bet thus became part of the Australian gambling landscape. In turn, more attention also started to be paid to relationships between more traditional forms of indigenous gambling (Brady 2004) and the emergence of commercial gambling industries in places like the Northern Territory and other regions of northern Australia (Young et al. 2007).

The forging of governable and taxable spaces for the activities of commercial actors is the point, I would ague, where we can start to talk of the emergence of cultures of gambling *consumption*. It is at this point that cultures of audit and standardization start to be applied to gambling commodities, ensuring levels of accountability and consumer fairness, and securing streams of government revenue. It is tempting to see such developments as simply an efficient means of extending participation in, and hence profiting from, socially embedded gambling forms such as horse racing. It is also tempting to view contemporary gambling consumption as a predominantly straightforward exchange of money for entertainment, with the 'interest' of gaining a windfall from gambling merely one of the attractions of this form of leisure. However, in this article I will emphasize a particular form of interaction within popular contemporary commercial gambling contexts. This is the interaction between knowledgeable and interested gambling actors and the various forms adopted by powerful commercial systems to manufacture and sell gambling commodities. I will suggest that the dialogic and interactive relationship between innovative commercial systems and gambling consumers has a profound influence on the instituting of cultures of gambling consumption. Commercial relationships and the industrial dynamics of rationalization, technological advance and commodity enhancement operate to reorder gambling culture to some extent. As we will see, this reordering may even spill over into the culture of the game.

It is not so much the decline of socially embedded cultures of gambling, such as that surrounding horse racing, that I want to describe (if indeed such a decline can be said to be taking place). Rather, I want to focus on the way gambling business and government regulation frame contemporary experiences and understandings of gambling as forms of mass consumption, but without eliminating *homo aleator* from the equation. The phenomenon of industrialized consumption of gambling products has seen gambling expenditure (losses) escalate to around A$15 billion annually. The relationships between this culture of mass consumption and socially embedded cultures of knowledgeable gambling are complex and as yet quite opaque. However, relatively well-informed gambling actors continue to operate in a variety of contemporary commercial gambling contexts. These actors must deploy knowledge and strategy in adapting to the dynamics of commercial systems that stage and frame gambling markets.

The paper proceeds by first briefly defining industrialized and commercialized gambling, and the framing of markets for the mass consumption of gambling products.

Examples of processes of adaptation by participants in three gambling consumption contexts are briefly outlined to give an idea of the diversity of spaces of gambling consumption. The main body of the paper then considers one example in more depth: the framing of sportsbetting markets for Australian Football League (AFL) matches. In particular, it looks at circumstances in pre-season and late regular season matches that call into question the integrity of the gambling product being offered to consumers, at least in the way it is theoretically framed. The case shows how the logics of commercial gambling and of football management can come into conflict. The discussion then highlights how knowledgeable actors are in a position to appreciate certain 'externalities' that might be influencing betting behaviour and markets. The concluding discussion draws on these examples to argue that the interaction of powerful commercial systems and knowledgeable actors is a key dynamic in contemporary cultures of gambling consumption. I finish by speculating that this dynamic will continue to shape transformations in gambling and its framing in social and cultural contexts.

Commercial systems, culture(s) of gambling consumption and adaptation

Western commercialized and industrialized gambling is perhaps best defined by Reith (1999, 1) as an activity that is 'strictly demarcated from the everyday world around it and within which chance is deliberately courted as a mechanism which governs a redistribution of wealth among players and a commercial interest or "house"'. The setting of rules and regulations, and the calibrating of technologies, govern the precise specification of 'house' advantage. The setting up of 'house' interests via the liberalizing of commercial gambling markets in Australian states and territories has been a form of coordinated economic action undertaken by coalitions of governments, state agencies, and corporate and other private-sector organizations (Livingstone and Woolley 2007). The introduction of casinos, hotel and club poker machine gambling and sportsbetting (including betting on virtually any independent third-party event such as elections) has achieved its aim of facilitating significant redistributions of wealth (stakes) as witnessed by burgeoning private profits and public revenues.

In Australia, the most recent survey (June 2005) showed a total of 5370 businesses involved in providing gambling, of which 77.8% were clubs, pubs, taverns and bars (ABS 2006, 5). In total consumption terms, commercial gambling expenditure was $16,910.3 million in 2004–2005. This equated to 3.05% of household disposable income that year. The dominant market was poker machine gambling in clubs and hotels, which accounted for 59.7% ($10,095.5 million) of total expenditure (gambler losses), compared with casinos (15.6%, or $2638.5 million), off-course wagering on racing (TAB) (11.5%, or $1936.5 million) and lotteries (9.6%, or $1619.1 million) (OESR 2006). State and territory governments received $4772 million in taxation revenue from gambling in 2006–2007, of which $2958 million came from poker machine gambling in clubs and hotels (ABS 2008).

The rapid institution of commercial activity at this scale requires substantial investment, pre-existing infrastructure such as telecommunications networks, and significant research and development (R&D) of technically based commodity forms around which transaction chains can be organized. At the same time, a regulatory regime, consisting of licensing processes, monitoring and auditing practices, and industry product standards, is required. Relatively new gambling sectors such as online sportsbetting are introduced 'off the shelf' through coordinated commercial and governmental action. Older sectors such as horse and dog race wagering have evolved more organically, from face-to-face interactions at racetracks and illegal off-course starting price (SP) bookmakers through a range of technical platforms that have changed the culture of wagering (e.g. the emergence of off-course totalisator wagering that eliminated the SP bookies). The advent of the Internet has provided a new medium for the sale of 'older' gambling products, while encouraging innovation in the product offerings available. Racing has adopted networked computerised information technology (CIT) based infrastructure over the past two decades, enabling expansion of both the channels of product delivery and the commodity forms that can be sold. For example, totalisator agencies now sell lottery-style 'quick pick' wagering products, with the selections of numbers (horses) done by random number generators (RNGs).

Variations in the evolution of gambling forms are of interest in and of themselves; however, the purpose of this brief outline is really to make one simple point: that powerful commercial systems have been put in place to enable gambling commodities to be staged for sale to an expanded market of ordinary consumers and committed gamblers alike. The regulatory, commercial and technical staging of gambling products for sale constitutes what I shall refer to as the 'framing' of gambling markets (Callon 1998). The process of framing creates spaces for new or transformed cultures of consumption of gambling products to emerge.

The framing of spaces of consumption of gambling is thus as diverse as gambling forms themselves. Hence quite varied processes of adaptation and learning can evolve in practice within them. Before going on to a detailed case study from the field of sportsbetting, I will first illustrate something of this variation via brief descriptions of two quite different framings of gambling activity, and of the kinds of adaptations that can emerge and evolve within these spaces of participation and consumption. The first example is the playing of the card game poker against other gamblers via online gambling websites. This form of gambling has become very popular throughout the world as part of the Internet revolution, while poker itself has developed something of a cult status in recent years, with competitions televised and local 'tournaments' being popular in Australian hotels and clubs. The online game is played in a virtual room in which remotely located gamblers compete against each other. However, what separates poker conducted around a table and that played via the Internet is the use of technical prostheses to assist play. Online poker players routinely use 'bots', small applications that monitor the betting behaviour of their competitors, to assist their gambling strategy. A gambler playing with the assistance of a 'bot' will know the betting patterns of others around the virtual table – for example, how aggressively they tend to bet the relative strength of their revealed hands. This creates an advantage for the 'bot'-assisted gambler, who can consider the behaviour of an opponent on a current hand in comparison to their historical pattern. These data add a stream of systematic information to their betting strategy.

The online poker player who uses a 'bot' thus has certain informational advantages over those who do not use this form of technical prosthesis to enhance their calculations. However, what has tended to occur in online poker games has been the adoption of 'bots' by the vast majority of players, particularly those who play seriously and/or regularly. In the absence of one form of information about an opponent – that is, their physical appearance, mannerisms or apparent emotional responses – online poker players substitute a system of data collection and processing that illuminates opponents' behaviour. The suspected use of 'bots' by most or all online poker players leads gamblers to adapt their behaviour to incorporate the role of 'bots' in the game strategy. For example, as one gambler described it, you have to be aware that some players will gamble in a seemingly reckless way on several hands in which they do not have good cards to disguise the situation when they do have good cards and want to bet big. These gamblers are playing to their opponents' 'bots', signalling through the historical data that are being compiled and interpreted about their behaviour that they do not need to have really good cards to significantly increase the size of their bets. The hope is to lull their opponents into thinking this is their typical pattern, making them more likely to take greater risks against this player on the occasions when the cards really have fallen their way. The knowledgeable agents of online poker games have thus had to adapt their strategizing to incorporate the capabilities of technical enhancements to individual poker-playing skills. This highlights the way in which the conduct of a particular game can be altered by technology and the qualities of the commercial system staging the gambling product. In this case, once the medium of the game is digital, new strategies and capabilities also come to the fore that transform gamblers' behaviour.

Technology is also important in the interactive relationship between poker machines and gamblers. Poker machine technology is big business, servicing a multi-billion dollar consumption sector in Australia. Innovation in the design and manufacture of poker machines is designed to extend the amount of time and money spent gambling. Extending gamblers' time on device (TOD) and the revenue returned per available customer (RevPAC) requires an engineering solution that is entertaining and satisfying in relation to the rate at which financial losses occur. The knowledge inputs that underpin poker machine technologies approach this task via two related strands of human science research. The first uses principles of behavioural psychology (Skinner 1953), in which the redistribution of gamblers' stakes as wins and losses occurs in such a way as to operate as a random ratio reinforcement schedule, disguising losses through the intermittent experience of machine events including wins, bonus features and free games. The second strand of research uses cognitive science to create the illusion of increased chances to win – for example by regularly featuring winning symbols (near-misses) or providing multiple lines to enable simultaneous bets and the illusion of more chances to win.

The capabilities, features and attractions of poker machine technology are not static, but rather are the outcome of processes of industrial R&D[1] and incremental innovation that lead to 'new' product offerings. The interactive relationship between poker machine technology and gamblers is thus always evolving. The strategies that gamblers adopt, despite the random nature of poker machines that means gamblers cannot influence game outcomes, are thus formatted and transformed by changes in the game technology. While it has been shown that many poker machine gamblers do not really understand that the outcomes of the game are entirely beyond their control (Productivity Commission 1999), it is important to realize that the play of all poker machine gamblers is to some extent configured in the interactive relationship with technology. While poker machines are not games of skill, the forms of behaviour and actions required by the obdurate materiality of poker machine technology provide the experience of interactive agency that engages gamblers in the activity. While contemporary poker machine gamblers who touch the screens and push the buttons on current networked CIT devices share a fundamental experience with those who pulled the handle of the old mechanical steppers that have now vanished from the market, the profusion of winning possibilities, game options and forms of reinforcement has led to an intensification of the interactive agency of poker machine gamblers. The relatively high level of consumption of poker machine gambling in contemporary Australia therefore cannot be understood adequately without taking into account this transformation of technology and the adaptation of gamblers to the experiential possibilities the advances in technology have presented.

The interactions between commercial systems and consumers in the two examples described are both technologically mediated. In the case of online poker, the adaptive behaviour of gambler is framed cognitively in terms of strategy and reflexivity. In the case of poker machine gamblers, adaptation to the commercial system is more practice based, with gambling behaviour incrementally reformatted over time by the new betting options and game features presented. To highlight the diverse 'framing' of gambling markets, I want to contrast these forms of adaptation with that of gamblers who bet on the outcome of AFL matches. In particular, I want to highlight the way in which these gamblers need to strategize in terms of seemingly 'external' factors, namely aspects of competition rules and logics of football management, which are not proper to the isolated sporting contest upon which bets are actually made, but rather are aspects of the institutional and commercial contexts that frame these contests and the AFL competition as a whole.

Gambling on AFL football

In 2005, the Australian Football League (AFL), responsible for the administration of the national Australian Rules Football competition (also known as the AFL), entered into commercial arrangements with the Melbourne-based Australian stock exchange-listed gambling company Tabcorp and the London-based betting exchange Betfair to provide betting services on its official competition matches. The agreements were for five years and gave the two companies exclusive rights to provide betting services on AFL-sanctioned matches in exchange for an annual fee estimated at around $1 million dollars for each company. The companies are also entitled to use the AFL brand in their marketing.

A range of betting options is now provided by the two companies – for example, on leading individual awards and end-of-season finishing order of the 16 AFL clubs. The major betting opportunities, however, are based on the outcomes and margins of the eight matches staged in each of the 22 rounds of the regular season, the nine matches of the finals series, and the pre-season knockout competition conducted in the month preceding the regular season, which commences in late April. Tabcorp offers totalisator and fixed-odds betting products, with totalisator bets entitling the holder to a proportional share of the total prize pool depending on the size of their wager relative to all those who made the correct selection. Betfair is a betting exchange, which matches the bets offered by clients with willing counterparties, enabling their clients to bet directly against each other while the company receives a commission for providing the 'matching' service. Large volumes of betting on AFL matches are also conducted via sportsbetting companies licensed and operated in the Northern Territory, via online and telephone accounts.

Interestingly, the AFL's commercial agreements with Tabcorp and Betfair are described as 'partnerships' or 'product and integrity arrangements'. The basis of these partnerships is a mutual recognition of the importance of the integrity of the product being offered. In other words, the independent third-party event (the football match) is considered to be of reduced value to both the AFL and the betting agencies if the integrity of outcomes is perceived to be corrupted or vulnerable to corruption. A crucial part of the commercial partnerships is thus an arrangement whereby the betting companies provide client data to the AFL, enabling the AFL to monitor participation in gambling on its sanctioned games.

The AFL gambling policy restricts footballers or other AFL employees from directly or indirectly betting on AFL-sanctioned matches or events contained in these matches. It is also contrary to the policy to assist others to gamble or to disclose information about players, tactics or other matters that are not already in the public domain. The key measure of the AFL gambling policy is thus to ensure the perception of credibly independent match outcomes by preventing footballers or other participants in the staging of its sanctioned matches from being involved in any way in betting markets staged on the outcomes of these matches. Since these arrangements were put in place, several AFL players, and at least one board member of an AFL club, have been sanctioned or warned for betting on AFL matches in contravention of the policy.

The integrity arrangements built into gambling business agreements are thus designed to frame the outcome of matches as depending entirely on proper 'footballing' factors. From the gambler's perspective, the equation is designed to be simple: which team is more likely to win and by how much? Football factors, which I will conceptualize as 'internal' to the framing of the market on a particular match, include home ground advantage, head-to-head records, weather or ground conditions and the selected teams of the opposing clubs. These variables can fluctuate considerably (e.g. dry or wet conditions) but can be known by all. Internal factors, such as perceptions of home ground advantage, have been found to impact on aggregate betting strategies in some circumstances (Schnytzer and Weinberg 2008).

Information about these 'internal' variables is thus fundamental to assessments of risk and to formulation of betting strategies. In recent years, the AFL has taken measures to enhance the reliability of such information. For example, the AFL limited the discretion of football clubs to change selected players at the last minute, or at least to ensure greater transparency about the 'odds' of particular players being able to participate. This may be due to the perceived disproportionate influence 'star' players can have on the outcome of matches. Sanctions for withdrawals of players after a certain deadline were introduced in an attempt to make the practice less attractive to clubs. While nominating a star player unlikely to play can have an influence on the way opposition teams may plan for the game, it also can affect the decision making of gamblers. The changes to team nomination procedures instigated by the AFL had the effect of standardizing the information available to football gamblers, regardless of whether or not this was directly intended as an 'integrity' measure in relation to gambling. Taking all this information into account, football betting should thus be a simple equation for gamblers: allocate relative significance to 'football factors' and make a reasoned decision. This process, however, leaves one further assumption unchallenged: that both teams are equally 'willing' to win.

AFL gambling controversies: Inside and outside the frame

But do AFL teams always have an equal 'interest' in winning? What external factors might contribute to an asymmetry in the distribution of 'willingness' to win a match without it destroying the 'integrity' of the product? And what kinds of information and calculations are required in these circumstances, such that a gambler might be considered a knowledgeable actor? And finally, what kinds of information sources might be most helpful in aiding gamblers adapt to contexts involving variables that, *de jure*, do not exist, having been helpfully bracketed out via 'product and integrity arrangements' and the AFL's rules and regulations?

Two controversies stand out in relation to gambling on AFL matches. The first is betting on the pre-season competition, which is sanctioned by the AFL and therefore framed as having 'integrity' from the gambling perspective. According to a football management logic, the value of the pre-season competition often lies elsewhere than in winning matches or the competition. Many coaches seem to value more highly the provision of competitive practice matches and an opportunity for tactical testing, for example. The second controversy relates to the structure of the AFL's National Draft system, whereby outstanding young talents are distributed to AFL clubs by virtue of a series of 'picks', starting from the club that finished last and proceeding in order to the winner of the AFL Premiership. The controversy here lies in the Draft's relation to what is colloquially known as 'tanking'. Tanking refers to circumstances where teams that have lost the chance for real success part-way through a season are perceived to be content to 'bottom out' so as to secure the best selections from the following year's Draft. I will discuss each of these controversies in turn, with a view to shedding some light on the questions posed above.

The AFL pre-season competition is a knockout competition, commencing in early February and finishing in late March, with the final played two weeks prior to the start of the premiership season proper. From a marketing point of view, the competition is important in building interest in the season ahead, and for the clubs that contest the final there is the excitement of a minor trophy, which can also have knock-on financial benefits in terms of boosting memberships and sponsorships. However, such benefits must be weighed against key football management objectives, such as having players fit and fresh for the long season ahead. Prolonged involvement in the pre-season competition increases the risk of untimely injuries and the exacerbation of the general fatigue that is always a factor during the premiership season. Purely 'football' logic, geared towards the objective of success in the September finals series, may imply only limited involvement in the pre-season competition.

Controversy regarding the pre-season competition and the 'rights' of gamblers came to the fore in 2008. In the first round of the competition, the Sydney Swans played Hawthorn at Aurora Stadium in Launceston. With two minutes remaining in the match and Sydney trailing by two points, Sydney coach Paul Roos made a substitution, sending one of his players into the forward line. According to an official overseeing the interchange of players on and off the ground, Roos gave instructions to the player, using words to the effect of 'Go forward, just don't kick a goal'. The official reported Roos' alleged comments to the AFL, which then appointed retired Justice John Winneke to lead an investigation into whether Roos' conduct amounted to 'match-fixing', punishable by a $100,000 fine and an indefinite suspension.

The CEO of the AFL, Andrew Demetriou, explained the rationale for the case related to its anti-gambling/match-fixing rules. Demetriou was widely quoted in the media on this point, saying that: 'The rules are put in place because there are people who actually bet on these games' (Herald Sun 2008). In turn, Roos called for betting to be banned on pre-season cup games. In describing his bemusement at the charge, he said: 'Andrew [Demetriou] explained it had something to do with the gambling clause, so just trying to get your head around that is a little bit hard ...'. As Roos, an experienced premiership coach, described it, there was an apparent conflict between the integrity of AFL pre-season matches as gambling products, and the obligations of coaches. Roos commented: 'I think it's a fine competition to play in ... but most coaches don't want to compromise their preparation for round one ... More and more focus is on round one because we're paid to be effective and win games in the regular season' (Tomarchio 2008). Roos became even more explicit in describing a variety of logics underpinning participation in the pre-season competition: 'As long as everyone takes it as what it is – a pre-season competition with *everyone having varying degrees of what their outcomes need to be*. And it seems more so than ever more coaches are saying, "Well, I'm focused on round one of the regular season"' (emphasis added) (Ormond 2008). In a later statement immediately prior to his investigation hearing, Roos clarified his view of the competition further, stating that 'over the six years that I have been involved with the competition (NAB Cup), we have used it as a feeder competition for our young guys, and we did the same this year' (Herald Sun 2008).

The situation that arose in the Roos match-fixing case was not at all surprising to football aficionados. A trip around some of the football journalism websites and blogs found a range of opinion, but little dissent from the proposition that not all clubs are equally interested in winning the pre-season competition. Many bloggers and commentators were quite clear about the way they constructed their understanding of the controversy: as a conflict between gambling integrity and football management.

> One could argue that all the gratuitous rule changes for the NAB Cup show that the AFL is not taking the competition entirely seriously either. If they were that invested in credibility they'd use normal rules for the NAB Cup and save the experimental rules for the Regional Challenge.
>
> As far as the gambling issue goes, anyone betting on the Swans to win a pre-season game in 2008 deserves to lose it all. (Posted by Zingiber on 5 March 2008, 9.52 a.m.)

> I used to reckon Blighty was a genius but now Roosey is ... how good a coach is he that he can get within two points in a game he is not trying to win! Somebody once told me that jockeys don't always try (not that i believe that) and some have even been suspended for it. I bet they wish they were as clever as roosey and only lost the race by a short half head instead of being 6 lengths back under a tight hold. Hard to give a bloke time for that. Like I said, genius! (Posted by rocco on 6 March 2008, 3.25 p.m.)

> Anyone who would bet on the NAB cup should seriously consider seeking gambling counselling. (Posted by Psi on 6 March 2008, 8.20 a.m.)[2]

Football journalists are no less circumspect in their assessment of the objectives of clubs in the pre-season cup. In an article from 2006, entitled 'Clubs Don't Care About Cup', Michael Horan described the mood in AFL clubs as follows: 'While every AFL coach who straps on a clipboard wants to win, there is no doubt winning isn't everything in February.' He went on to state that 'most fans understand and accept' this situation. Interestingly, Paul Roos featured amongst those quoted by the article, some two years prior to his eventual 'trial', saying: 'Our goal is to prepare for round one and to give our younger players an opportunity to play against some quality opposition.' Despite the Sydney Swans being the 2005 AFL Premiers, bookmakers rated them as rank outsiders at $41 to win the 2006 pre-season competition. It would seem that bookmakers were therefore 'pricing in' certain externalities relating specifically to the pre-season cup and the relative 'willingness' of clubs to win matches.

The AFL investigation of the 2008 controversy found that Roos' comments did not lead to a conclusion that Roos had intentionally induced or encouraged a player to perform otherwise than on his merits. The AFL's Legal and Business Affairs manager, Andrew Dillon, commented that: 'The integrity of the competition is paramount and the AFL was obligated to fully investigate the concerns raised by match officials.' From the perspective of the core constituency of the AFL – football fans, the media and those involved in the clubs themselves – the result was entirely predictable. It appears clearly understood by the Australian Rules 'community' that winning is not the priority of many, if not most, clubs involved in pre-season matches. The pre-season has a particular place in football culture and, as the comments above reflect, there is little sympathy for those betting on the outcomes without an understanding of this broader context.

In fact, the AFL's assertion of the rights of gamblers to 'product integrity' as equivalent to the imperatives of those involved in football management were not viewed kindly by many insiders. These views were articulated by champion ex-player and then Brisbane Lions coach Leigh Matthews a month after the Roos investigation, which he described as ridiculous. Matthews' opinion on the situation was widely quoted: 'I think the AFL should say, "You want to bet on the footy fine, don't include us. Don't ask us to have rules and regulations that pander to people who might want to bet on the game."'

Matthews said he 'hated' the perception that imperatives arising from involvement in commercial gambling arrangements were a factor in the way football was run: 'The one thing that really annoys me is any thought that whatever we do as a game has something to do with people betting on it.'

The AFL's response to Matthews was to point out that its gambling integrity partnerships enabled it to monitor gambling on football. In this, the AFL might be accused of being a trifle disingenuous. Whilst no one 'inside' football culture seemed to be criticizing the procedures to monitor corruption through betting on football, an underlying resentment seemed to be expressed by Matthews, which was more to do with the corruption of football culture, broadly understood. Specifically in this case, dissatisfaction appeared to lie with the subjecting of the accepted function of the pre-season to criteria imposed by the framing of the competition as a gambling market. What Matthews, Roos, fans, journalists and other 'insiders' really seemed to be saying was that the AFL's investigation was largely rhetorical, designed to conceal the acknowledged asymmetry that exists in the willingness to win pre-season games. The AFL's interest in so doing appeared to be widely perceived as attempting to shore up the credibility of the gambling product, potentially at the expense of traditional football culture.

A second controversy surrounding gambling on AFL-sanctioned matches, which relates to the National Draft system of allocating new players to clubs, has become an annual site of conspiracy theories and intrigue which, whether or not 'real', has concrete repercussions for betting markets and gambling behaviour. The AFL National Draft is conducted in reverse of the finishing order of the previous premiership season, with the last-placed club having first choice amongst the new draftees and so on. In recent years, clubs that have won fewer than five games for two successive seasons have also been eligible for a priority pick of the draftees, a pick that occurs before all others. So, for example, although Richmond Football Club finished last in season 2007, its first pick of the Draft actually occurred after the priority pick 'earned' by Carlton Football Club, which finished above Richmond but had won fewer than five games in 2006 and 2007. The controversy associated with access to the player Draft is known colloquially as 'tanking', which can be summarized as endeavouring not to win too many games, either relative to other clubs or to thresholds for priority picks, such that opportunities to profit from the Draft would be reduced for little apparent gain.

Tanking is considered by AFL watchers to be more subtle than merely 'trying to lose'. This is because football management strategies focused on longer term outcomes may nevertheless influence short-term results. All AFL clubs have a limited number of players permitted on their 'playing list', so strategies to shape the list are strategic and medium term. List management is thus an important part of clubs' longer term performance objectives. Where a club looks unlikely to participate in the finals series in a particular season, list management decisions such as sending star players for rehabilitating surgery or blooding young players to accelerate the gaining of experience, coupled with match-day experimenting with player positioning or team tactics, are arguably sound longer term football management strategies. These list management and game-day decisions may therefore be beneficial to the future of the club, but have short-term costs in terms of match results. List management 'tanking' over time, so the theory goes, may also bring the indirect benefit of earlier selections in the National Draft than would otherwise have occurred. Tanking by deliberately 'throwing' matches is thus not considered to be the key element of the tanking controversy.

'Tanking' is also a mediatized phenomenon. For example, following a disappointing first half of the season, Port Adelaide coach Mark Williams described his strategy for the remainder of 2008 as 'picking for the future', leading to headlines that he had 'admitted' to tanking. Various ex-players and officials have described the list management practices of their own or other clubs as effectively 'tanking the season', with some saying that it was difficult to argue that winning was in the best interests of clubs at certain times. One former AFL player and senior coach summarized tanking as a structural problem that arises by mid-season every year:

> How times have changed. Where once teams had been highly motivated to avoid the ignominy of the wooden spoon, now we have a system that makes it a tantalising object of desire ... By the mid-season break the 16 teams that start the season full of anticipation and excitement become more like 10 to 12 that still have finals hopes. Four to six teams already have made a decision they cannot make the finals, so are duty bound to look after the best interests of their club. They effectively have 'put the flag up'. That can only mean the draft for the following season and this is where a problem is emerging. Clubs are tanking the season. (Thomas 2007)

The continual discussion and debate about 'tanking' in the media is also important in relation to the 'credibility' of AFL-sanctioned matches. One 2008 AFL coach described 'tanking' as a 'slur' on the game. Another described the perception of tanking as itself a matter of concern for the AFL administrators:

> Whether there is tanking or not, and I don't think anyone can really assess that, I don't think it's that clear-cut, I think it's more a perception. And if the perception's out there ... you'd certainly think the AFL could probably look at it. (The Age 2008)

The widely perceived and mediatized controversy over 'tanking', mythical or otherwise, can be considered an epiphenomenon of institutional policies and regulations that are nominally 'external' to the outcomes of sanctioned AFL matches. The National Draft is designed to create a process which, all other things being equal, would lead to an even distribution of playing talent and hence on-field success. This is far from the reality, of course, but 'tanking' refers to the perception that seeking to accelerate the process of performance decline, and thereby the harvesting of new talent, can lead to a more rapid rejuvenation of on-field fortunes than would otherwise have occurred. Cases cited in support of this argument include St Kilda and Hawthorn, which both 'bottomed out' and drafted large numbers of highly regarded playing talent. Hawthorn rebounded from 15th in 2004 and 14th in 2005 to Premiers in 2008, largely due to wise selections using early selections among Draft talent. St Kilda went from last in 2000 and second last in 2002 to top four finishes in the period 2004–2006, but, unlike Hawthorn, is widely perceived to have 'missed the window of opportunity' created by its prior 'bottoming out'.

There is no evidence that Draft considerations alter the probabilities of struggling clubs winning matches (Borland 2006). However, real or imagined, tanking does create controversy around betting on AFL matches. Late in the season in particular, gamblers who bet 'objectively' on factors 'inside' the frame of the AFL betting product – that is, without knowledge of the draft system and its implications – would seem to be at a disadvantage. This is perhaps best illustrated through an example. The final regular season round of 2007 featured a match between the 14th-placed (of 16) Melbourne Football Club and the 15th-placed Carlton Football Club. Despite a series of objective football factors – the clubs were closely matched in overall performance; Melbourne were missing several key players; Melbourne had suffered several heavy losses in recent matches – that would normally indicate both clubs had a reasonable chance of winning the match, bookmakers set opening prices that suggested this was not the case. A rough survey of the prices being offered by bookmakers on the match showed Melbourne at around $1.25 and Carlton at more than $3.00.

The unusual calculation of these odds was well understood in football circles and prominent in the print and television media. One bookmaker was widely reported as delaying the setting of odds on the game and handling it with 'absolute kid gloves'. As this bookmaker saw it: 'The odds should be a lot closer than what they are but because of the widespread talk about Carlton wanting to lose they're not' (The Age 2007). Another bookmaker reported taking bets of $20,000 on the match, stating this was far more than was usual for a 'dead rubber' at the end of the season. Carlton had won just four matches in 2006 and only another four up to the final game of 2007. If the team lost, it 'earned' a priority pick to be taken before all others in the National Draft; along with its pick for finishing 15th, it would effectively have picks one and three. Alternatively, if the team won, Carlton would end up with only the third pick overall. Melbourne had also only won four matches in 2007 prior to the last round, and by losing stood to 'earn' a priority pick at the start of the second round of the draft, in effect pick number 17. Both teams appeared to have far more to gain by losing than by winning this otherwise meaningless match. As one political journalist described it: 'All that has to happen now is for Melbourne and Carlton to somehow find a credible way not to win' (Happell 2007).

In the end, Carlton lost the game and benefited from first and third picks in the 2007 National Draft. Early in 2008, a former Carlton assistant coach, Tony Liberatore, who had been part of the match-day coaching team at Carlton for 2007, including the final match, discussed the tanking issue on Channel Nine's The Footy Show. In the course of this discussion, Liberatore said that winning matches 'wasn't the be all and end all'. Describing the match-day situation, he said that he had 'never heard [a directive to lose], but I could feel it, if that makes sense. Nobody ever said we're not going to win but the feeling in the group was that it was a bit of a laugh' (ABC 2008). The AFL voiced concern about the statements, but after meeting with Liberatore terminated any further investigation of the matter.

Conclusion

The AFL's commercial partnerships with gambling companies oblige it to deploy a variety of strategies to ensure the integrity of its sanctioned matches as betting products. Despite such strategies, the perception that tanking occurs in the AFL has implications for gamblers and gambling providers, altering the behaviour found in betting markets. The cultural knowledge and adaptive behaviour of bookmakers and punters are crucial to the functioning of these markets. As we have seen, bookmakers do not set odds on matches solely on the basis of the framing imposed by the AFL, which would mandate a symmetrical distribution of willingness to win a particular match. Rather, bookmakers 'price in' a range of 'externalities'. In the case of 'tanking', this means precisely the perception that there is an asymmetric distribution of willingness to win, due to the influence of the National Draft process.

Viewed from inside the frame, the integrity of gambling on third-party events such as sporting competitions largely depends on the soundness of the information provided. However, gamblers who rely on such 'objective' information streams would appear to be at a disadvantage compared with those who have 'grassroots' knowledge of sports cultures and institutional factors that, nominally at least, are bracketed off from game outcomes. Knowledgeable gamblers in such circumstances are those who understand the contextual factors that need to be dragged into the frame, alongside results, players, form, and so on, and can therefore understand both the externalities that bookmakers are 'pricing in' and whether the prices they set in response to this wider set of factors represent an attractive premium on the risks associated with betting on such controversial matches.

As we have seen by looking at gambling controversies surrounding pre-season and late-season games in AFL-sanctioned competitions, such cultural knowledge is invaluable, with a range of informal information sources, such a blogs, Internet forums and the mainstream media constituting important resources. Of course, without the appropriate prior processes of acculturation, sifting through informal and mediatized representations of likely future risk events is likely to be as confusing as it is enlightening. Insiders, embedded as they are in cultural contexts, are much more likely to be able to navigate the complexity of gambling markets such as that for Australian Rules Football. Such adaptive behaviour among gamblers is, in itself, not new. However, what the example discussed highlights is that the commercial arrangements that frame sports competitions as gambling markets do not prevent the spillover of factors that were intended to be excluded from the gambling space. The ongoing evolution of gambling culture thus includes adapting to interactions shaped by powerful commercial actors and systems, both from inside and outside the frame.

In conclusion, the commercial liberalization and industrialization of gambling underpins a dynamic of innovation, ensuring the emergence of new products and forms of staging and delivering gambling commodities to consumers. Commercial gambling, whether playing cards online, interacting with new poker machine technologies or grappling with the imposed framing of sports betting, changes gambling culture. Sometimes change is quite incremental and at other times quite dramatic and radical, as we have seen with the emergence of CITs and networked forms. However, as we have also seen, gamblers and embedded gambling cultures are knowledgeable, resilient and adaptable. In Australia, contemporary cultures of gambling consumption are emerging via dynamic relationships between commercial and institutional actors, knowledgeable gamblers with their accumulated cultural resources and broader pools of consumers. This process is never static or complete; rather, the flow of innovation and adaptation in gambling is restless, ceaseless and transforming. The qualities of gambling commodities and gamblers' experiences on offer today are not necessarily those that will characterize tomorrow.

Notes

1. Aristrocrat P/L reported spending A$104.2 million on R&D in 2006–2007.
2. Comments posted on http://blogs.theage.com.au/realfooty in response to the unattributed article entitled 'Credibility and the Cup'.

References

ABC (Australian Broadcasting Commission). 2008. Liberatore to meet with AFL over tanking claims. Unattributed article. 14 March. http://www.abc.net.au
ABS (Australian Bureau of Statistics). 2006. *Gambling services*. Cat. no. 8684.0. http://www.abs.gov.au

———. 2008. *Taxation revenue Australia 2006–07*. Cat. no. 5506.0. http://www.abs.gov.au

The Age. 2007. AFL bookies react to tanking debate. Unattributed article. 28 August. http://www.theage.com.au

———. 2008. Eade joins call for AFL tanking action. Unattributed article. 18 June. http://www.theage.com.au

Basu, E.O. 1991. Profit loss and fate: The entrepreneurial ethic and the practice of gambling in an overseas Chinese community. *Modern China* 17, no. 2: 227–59.

Borland, J. 2006. Economic design and professional sporting competitions. *Australian Economic Review* 39, no. 4: 439–41.

Brady, M. 2004. *Regulating social problems: The pokies, the Productivity Commission and an Aboriginal community*. CAEPR Discussion Paper no. 269. Canberra: Centre for Aboriginal Economic Policy Research, Australian National University.

Callon, M. 1998. An essay on framing and overflowing: Economic externalities revisited by sociology. In *The laws of the markets*, ed. M. Callon, 244–69. Oxford: Blackwell.

Charlton, P. 1987. *Two flies up a wall: The Australian passion for gambling*. Sydney: Methuen Haynes.

Happell, C. 2007. Still plenty in the tank for AFL losers. 27 August. http://www.crikey.com.au

Herald Sun. 2008. Paul Roos would like to see pre-season cup betting banned. Unattributed article. 5 March. http://www.news.com.au/heraldsun

Horan, M. 2006. Clubs don't care about the cup. *Herald Sun*, 23 February.

Livingstone, C., and R. Woolley. 2007. Risky business: A few provocations on the regulation of electronic gaming machines. *International Gambling Studies* 7, no. 3: 343–58.

Lynch, R. 1990. Working class luck and vocabularies of hope among regular players of the poker machines. In *Sport and leisure: Trends in Australian popular culture*, ed. D. Rowe and G. Lawrence, 156–78. Sydney: Harcourt Brace Jovanovich.

McMillen, J., ed. 1996. *Gambling cultures: Studies in history and interpretation*. London: Routledge.

O'Hara, J. 1988. *A mug's game: A history of gaming and betting in Australia*. Sydney: University of New South Wales Press.

OESR (Office of Economic and Statistical Research). 2006. *Australian gambling statistics 2005–06*. Brisbane: OESR.

Ormond, A. 2008. Ban Cup betting: Roos. 4 March. http://afl.com.au

Productivity Commission. 1999. *Australia's gambling industries*. 3 vols. Canberra: Ausinfo.

Reith, G. 1999. *The age of chance: Gambling in Western culture*. London: Routledge.

Schnytzer, A., and G. Weinberg. 2008. Testing for home team and favorite biases in the Australian Rules Football fixed-odds and point spread betting markets. *Journal of Sports Economics* 9, no. 2: 173–90.

Skinner, B.F. 1953. *Science and human behaviour*. New York: Free Press.

Thomas, G. 2007. Tank is weapon of choice in basement battle. 8 July. http://www.realfooty.com.au

Tomarchio, C. 2008. Paul Roos to face AFL investigation. 4 March. http://news.theage.com.au

Young, M., T. Barnes, M. Stevens, M. Paterson, and M. Morris. 2007. The changing landscape of Indigenous gambling in northern Australia: Current knowledge and future directions. *International Gambling Studies* 7, no. 3: 327–43.

Localizing a global amusement park: Hong Kong Disneyland

Anthony Fung and Micky Lee

Since the 1990s, scholars have considered the implications of global media on local cultures. Most scholars would agree that transnational media corporations disseminate and promote their cultural values through products and services. However, there is little consensus among scholars about whether transnational corporations homogenize local cultures and values. Unlike scholars, transnational corporations take a more pragmatic view of globalization: to them, globalization and localization are strategies to maximize profits. Regular readers of the *Wall Street Journal* learn that transnational corporations do not hesitate to localize products and business practices in foreign markets. For example, McDonald's has different menus in different markets to suit local tastes (Ritzer 2008). At the same time, businesses in local markets appropriate practices of transnational corporations. For example, eateries adopt McDonald's-like fast food operations to serve traditional Chinese food in Shanghai. Chan and Ma (2002) call this a two-way transculturation process, in which one culture is transformed by another and vice versa. In the process of hybridization, local culture is transformed into global culture, and global culture becomes part of local culture.

No company represents the conquering cultural army better than that of Walt Disney. Since the founding of the Disney Company in the 1930s, Mickey Mouse and his cohorts have been symbols of American imperialism and cultural hegemony, and more recently of globalization. Prior to the opening of Euro Disneyland, Ariane Mnouchkine, a French theatre director, called the park a cultural Chernobyl. Anti-globalization activists target the Disney Store for protests as much as they do McDonald's and Starbucks. It is indisputable that the Disney Company is a powerful cultural and political economic institution. However, unlike other transnational companies, Disney seems to be oblivious to local tastes in foreign markets. While transnational corporations such as McDonald's, Nike and Wal-Mart try to accommodate consumers' tastes in foreign markets, Disney hardly modifies its cultural products for non-US consumers. Why does Disney refuse to localize its products in different markets? Given Disney's hegemonic power, why would local consumers be willing to subject themselves to cultural homogenization by a foreign entity?

To answer these questions, we focus on Hong Kong Disneyland and argue that the concept of 'localization', used in both academic and business discourse, is too rudimentary to capture the complex cultural adaptation of Disneyland in China. The concepts of 'globalization' and 'localization' assume that culture and value are static entities and that there is a well-defined boundary between the global and the local. The concept of 'local' implies a physical place where a homogeneous group of people shares the same culture and the same values. The concept of localization is particularly problematic in the context of global cities (as discussed in Sassen 2001). A global city is defined here as a strategic site where global capital accumulates; this site attracts both highly specialized professionals and an unskilled global underclass. Who are the locals in a racially and economically diverse global city? What are local cultures and values in global cities if those cities were colonized and occupied by foreign forces? Although transnational corporations are the most easily recognized agents that impose manufactured, commercial cultures on consumers in different markets (Gerhson 1997), when culture is transmitted or imposed the history of colonization and decolonization, socialism and capitalism should not be discounted. Hong Kong is a global city that was a British colony for 150 years. Currently, as a Special Administrative Region of China, Hong Kong's executive, legislative and judiciary systems remain independent of the People's Republic of China until 2047. Whereas the PRC brands itself as a socialist capitalist country, the Hong Kong market is one of the freest in the world. Disney's choice of Hong Kong as the site for the first Disneyland in China is strategically sound.

Disney as the American way of life?

Scholars and cultural critics have long argued that Disney exercises cultural and political imperialism to spread American culture and ideals to the rest of the world (the most remarkable criticism being Dorfman and Mattelart 1991). Morley and Robins (1995) go to the extent of calling Disney a 'shared culture' that is universally manufactured and distributed. They argue that globalization is an international reconstruction of a global–local nexus. As Western values are disseminated through global capital, a new relation between global and local spaces is formed. To them, global culture does not erase local cultures, but the global does overshadow the local.

The Global Disney Audiences Project by Wasko, Phillips, and Meehan (2001) confirms that Disney products do have the power to influence the impression that audiences have of Disney. Audiences from all continents overwhelmingly agree that Disney values are about family, fantasy, fun, good over evil, happiness, imagination, love/romance and magic. The aforementioned project also shows that most of the audiences have seen at least one Disney film.

Unlike most popular global brands such as McDonald's, Coca-Cola and Pepsi, Disney fully exploits a strategy of diversification. The Disney brand name is sold through different media and venues, including film, television, books, magazines, merchandize, ice-skating shows, musicals, amusement parks, and so on. Like the makers of other global brands, Disney implements a strategy to 'catch the consumers when they are young' by targeting children. According to the Global Disney Audiences Project, the average age for an audience member's first contact with Disney products is 4.6 years. Audiences also self-reported that they were most fond of Disney when they were between 1 and 12 years of age. Disney values are internalized by consumers during childhood.

Hong Kong children who live on the periphery of the global economy (McPhail 2002) learn of Disney through both official and unofficial outlets. Like American children, Hong Kong children experience the Disney brand through films, television, children's books, the Disney Store and now Hong Kong Disneyland. In addition, non-licensed and counterfeit Disney products are readily available in gift stalls, from hawkers and at wet markets. Scholars such as Friedman (1995) and Pieterse (1995) call this the hybridization and creolization of cultures: Disney characters such as Winnie the Pooh, Mickey Mouse and The Incredibles appear on a wide array of locally made children's products (e.g. towels, silverware, water bottles, stationery) that are not available in the United States. In the same vein, the Disney character Mulan is a cultural hybrid that speaks to both Chinese and non-Chinese audiences (Chan 2002).

Local bootleggers may juxtapose the six princesses of Disney – Jasmine, Cinderella, Sleeping Beauty, Snow White, Ariel and Belle – on one product. Other bootleggers modify Disney characters just enough to avoid copyright infringement. Despite Disney's aggressive lawsuits against copyright violators, locally made bootleg merchandise may enhance Disney's influence and increase profits. Daya Thussu (2000) indicates that global media facilitate a new consumer lifestyle, and this encourages locals to be more receptive to global products. The Disney products that parents in Hong Kong give to their children may be counterfeit, but such non-licensed products may nonetheless enhance the children's enjoyment of Hong Kong Disneyland.

Some scholars (e.g. Lull 1995) argue against the power of globalization to erase local cultures. These scholars believe that cultural texts are polysemic – that is, the texts are open to multiple interpretations. To refer to the Global Disney Audiences Project again, audiences may have similar understandings of Disney, but the contexts in which they arrive at this understanding are always culturally specific (see Reis 2001 on Mexico, and Yoshimi 2001 on Japan). Winseck and Pike (2007) state that the extent to which American culture has globalized local cultures is debatable.

Implicit in the scholarly discussion of the relation between global and local cultures is the notion that local culture is always more authentic and more valuable to the population than global culture. As mentioned previously, the flow of cultures has to be understood in a historical context. In socialist countries such as the PRC, consumerist, manufactured global culture may evoke 'implacable contradictions existing in the popular consciousness' (Lull 1995, 260). An unintended consequence of the spread of global consumer culture could be that the citizens of a strictly controlled socialist state may develop aspirations for more personal freedom and greater rights, which in turn could undermine the legitimacy of the socialist regime.

A methodological note: Ethnography

Since Hong Kong Disneyland opened on 12 September 2005, the two authors have paid several visits to the amusement park with their families and friends. Anthony Fung and Micky Lee grew up in Hong Kong in the 1960s and 1970s, a time when Hong Kong's economy developed rapidly under the British colonial government. Like many Hong Kong academics, both authors studied for their PhD in the United States – Fung at the University of Minnesota and Lee at the University of Oregon. Their experiences as graduate students and US residents have helped them to critically assess the flux of local Hong Kong culture in the global economy. Since childhood, both Fung and Lee have been exposed to Disney through films, television, comic books, magazines and merchandise, and both were already familiar with Disney's parks when Hong Kong Disneyland opened: Fung had visited Disneyland in Tokyo and Paris before the Hong Kong park opened, and Lee was familiar with Disneyland from its presentation in books and television programs.

Fung visited the Hong Kong park with his family and different groups of friends, and Lee visited the park with her mother. Hence, we were able to observe how three generations of Hong Kong Chinese react to the park. Lee, who no longer resides in Hong Kong, visited the park once. Fung visited Hong Kong Disneyland during special festivals such as Christmas, New Year and Chinese Lunar New Year, and at other times as well. He spent a night and dined at the Hong Kong Disney Hotel – a replica of the Disney hotel in Florida. He attended a 'fairytale' Disney-themed wedding at the Disney Hollywood Hotel and dined there as well.

We conducted our 'fieldwork' by observing how the park's rides and other amusements are designed, arranged and promoted. We also paid attention to how the characters dressed and interacted with visitors. We noted the comments made by our family and friends about the park. During the visits, we 'felt' and 'tried' the 'localization' strategies of Disneyland, and we assessed whether the strategies worked for us as consumers. For the purpose of triangulation, we spoke informally to visitors of different backgrounds, and documented interactions between Disneyland staff and customers.

One Disneyland, three languages

Hong Kong Disneyland is a joint venture between the government of Hong Kong and the Disney Company. Although Disney paid for only 10% of the construction costs, it owns 43% of the shares in the park and has full control of the management.[1] Disney maintains strict control over the operation of Disneyland (Wasko 2001). The government of Hong Kong suggested that Chinese culture could be incorporated into the park, but the Disney Company argued that visitors should have an authentic Disney experience (Slater 1999). The Disney Company sees the Hong Kong park as a sample of the 'real, authentic' parks in the United States; it hopes Asian tourists will visit the US parks after having a taste of one in Hong Kong (Orwall 1999). Local visitors and the press lament the small scale of Hong Kong Disneyland; the global consumers see it as a reduced version of the California Disneyland: there is a miniature Sleeping Beauty Castle, a short Main Street USA, and a downsized Fantasyland.

Hong Kong Disneyland targets three groups of visitors: local visitors, mainland Chinese visitors and other Asian visitors ('Disney is optimistic' 2003; Hilsenrath and Coleman 1999). The Disney Company believes that the park attracts two kinds of tourists: 'induced tourists' who visit Hong Kong primarily because of Disneyland; and 'base tourists' who visit Hong Kong and make a trip to the park (Hilsenrath and Coleman 1999). The Disney Company is most interested in attracting mainland Chinese tourists, who would be unlikely to visit any Disneyland park if one were not located in Hong Kong. Like other transnational media corporations, Disney keeps a close eye on the mainland Chinese market. However, the PRC's strict investment and ownership regulations act to slow Disney's expansion into the Chinese media market. In addition, pirated DVDs and counterfeit merchandise deprive Disney of profits in what is potentially the largest market in the world (Fowler 2006). Hence the company views Hong Kong Disneyland as an opportunity to familiarize mainland Chinese with the world of Disneyland before the PRC relaxes regulations on investment and ownership.

Since three types of visitors are targeted, the park must offer trilingual services: Cantonese for local Hong Kong Chinese; Putonghua/Mandarin for mainland Chinese and

Taiwanese; and English for all other visitors. Signs in the park and visitor information guides include both English and Chinese text; the Chinese text is presented in both traditional Chinese characters for Taiwanese visitors and Hong Kong locals and in simplified characters for mainland Chinese. Performers, dancers and customer service personnel are largely local residents, but they are required to speak Cantonese, Putonghua and English. The existence of trilingual services unintentionally creates differences between tourists groups, dividing the global consumers from the aspiring global consumers. Among the local visitors, many are families who purchase annual passes so that they may pay regular visits; other locals visit the park to celebrate special occasions and festivals. The English-speaking tourists come from other parts of Asia and the world. The largest cluster of visitors is Putonghua-speaking mainland Chinese. Compared with the Cantonese- and English-speaking visitors, visitors from mainland China know little about Disney, and even if they recognize some Disney icons, the mainland Chinese tourists have little familiarity or affinity with the stories and values associated with Disney characters.

To a certain extent, the Cantonese- and English-speaking audience are global consumers exposed to homogeneous, commercialized American culture through multifarious channels, and as a side-effect of globalization they are inundated with entertainment that targets the consumer. The Putonghua-speaking audience, however, is less familiar with global culture as a consequence of the PRC's closed political system and imposed ideology. When mainland Chinese make the pilgrimage to Hong Kong and the Hong Kong Disneyland, they want to sample global culture and perhaps even embrace it, if only briefly.

Although there are hints of local Chinese culture in Hong Kong Disneyland's restaurants and shops, these cultural elements are intended primarily to sell Disney merchandise. For example, a Chinese restaurant in the park offers a variety of dim sum shaped like Mickey Mouse's head, and a jewellery shop sells figurines of Disney characters made of 24 carat gold. Faced with unsatisfactory attendance, the park management agreed to incorporate more Chinese cultural elements: Mickey Mouse wears a Chinese long robe; visitors are greeted by a character representing the traditional Chinese God of Wealth; and dragon dancing during the Chinese New Year (Fowler 2008).

Overall, the park and the two hotels are designed to reflect genuine Disney culture: visitors are encouraged to believe that they are not in the city of Hong Kong, but rather in the World of Disney. The park is built in a remote part of Hong Kong from which no skyscraper can be seen. When subway passengers take the Disneyland train, the environment of the train compartment reinforces the notion that visitors are no longer in the subway system: there are no advertisements suggesting a connection with the outside world, and the window is in the shape of Mickey Mouse's head. The railway station in Disneyland looks nothing like the modern subway stations in Hong Kong; with its cast iron stairs, the Disneyland stop resembles an old-time railway station. To reach the main entrance of the park, visitors must walk through a park that features a Mickey Mouse water fountain as its centrepiece.

Ironically, the railway station and the park with the Mickey Mouse fountain together occupy an area that is disproportionately large when compared with the size of the actual park. However, the size of this space helps visitors mentally prepare themselves to enter a magical world. For those who cannot afford to enter the park, the train ride and a stroll outside the entrance are like a free *amuse bouche* that heighten anticipation. For those like Lee's father who have no interest in the park, the train ride and the stroll still hold value as a trip *to* Disneyland.

Disney is ready to provide a 'genuine' Disney experience for visitors who are interested in Disneyland and who can afford to spend a night at the hotel. Every hotel room offers Disney-logo bedding, slippers, soap and towel set. Early in the morning, Disney princesses and other characters visit the lobby and take photos with children. Princesses such as Sleeping Beauty, Belle and Cinderella are played by Caucasians who can only speak English despite the presence of very young Chinese children. The characters Mickey Mouse and Buzz Lightyear do not speak Cantonese or Mandarin, but instead simply wave to children. Although most visitors would be familiar with the story of Mulan that existed before the Disney version, the character would not be found wandering in the lobby.

The world of Disney presented on the official Hong Kong Disneyland website shows the white princesses Snow White and Sleeping Beauty being admired by Chinese visitors. Disneyland recognizes that the Chinese visitors want a Western, global experience and fantasy, not a traditional or local one.

As demonstrated by the park, the hotel and the website, Disneyland presents itself as a pure experience untainted by Chinese culture. Disney did not conceive of the park as Hong Kong Disneyland, but rather as Disneyland in Hong Kong. Its goal is to train visitors to be global consumers who understand Disney and know how to enjoy it. The visitors may not be able to converse with the Disney princesses in English, yet it is precisely this 'foreignness' that encourages visitors to believe they are not in Hong Kong. In fact, hotel guests might find a Cantonese-speaking Chinese Sleeping Beauty extremely odd. Nevertheless, there are games in the park where language and Disney cultural knowledge are essential. In the following section, we analyse the global discrepancies highlighted by Stitch Encounter, Adventureland, The Festival of Lion King, the Winnie the Pooh ride and the Golden Mickey Show.

Stitch Encounter: Highlighting the global discrepancy

Stitch Encounter and the boat ride in Adventureland are two experiences in which visitors fully interact with Disney staff. For either ride, visitors can choose a session in Cantonese, English or Putonghua. Stitch Encounter is drawn from the *Lilo and Stitch* film and television program. The audience is expected to be familiar with the character and the story before taking part, as there is no explanation of who Stitch is and why he is in a spaceship. In Stitch Encounter, audiences are asked to sit in front of the screen that displays the control panel of the spaceship of which Stitch is the Captain. Children are invited to sit on the first row to chat with Stitch. In the first scene, as in the television series, Stitch commits a mischievous act by stealing a spaceship and flying across the galaxy.

In the Putonghua session, Stitch appeared friendly, but he liked to tease and play jokes on the audience. He spoke quickly, had a short temper, and even teased participants about their Chinese accents. In one of the sessions that Fung attended, Stitch's facial expression scared some young children and even made one child cry. In another session, Stitch's speech included pseudo-scientific words related to computers and space travel; the jargon may be incomprehensible to an audience that does not understand the purpose of the entertainment.

In the introduction to one session, Stitch pointed to a mainland Chinese participant and asked for his name and hometown. The participant naturally growled out an answer in front of the sizeable audience. Stitch then taunted the participant and the audience by repeating the man's words in broken Putonghua with a funny 'space alien' accent, adding that he had never heard of such a name and place. This conscious attempt to poke fun at the audience's background helped establish a distance between the fantasy world of Disney and the daily realities of the Chinese visitors. Stitch also picked a young woman from the audience and begged her to be his girlfriend, behaviour that led to another kind of culture shock for audience members. In summary, the interaction between the Stitch character and the audience reinforced the discrepancy between the daily life and culture of the mainland Chinese audience and the fantasy and manufactured global culture of Disney.

The boat ride in Adventureland served a similar function to Stitch Encounter. The Adventureland concept relies heavily on the First World fascination with the other. Despite decolonization and a heightened awareness of racism, Disney refuses to change the outdated view of Adventureland; instead, it extends its worldview to the other. The boat captains welcome Cantonese, Putonghua and English-speaking tourists aboard the *Congo Queen* and other boats with similar exotic names. Like Stitch, the boat caption is a con-archaeologist who acts as if he or she knew much more about the world than the passengers. Although the captain and some visitors share similar cultural and linguistic backgrounds, the caption acts as if he/she came from a part of the world where no human being dares to go. At one point, the captain pointed at the artificial Saharan animals, telling the passengers how to differentiate a female from a male zebra. When Lee's mother told the captain that the animals were mechanical displays, the captain rejected the reality by saying the animals were real. When Stitch engaged in the audience in a pseudo-scientific world, the boat captain plainly retold outdated myths about the Third World (such as barbarianism, and dangerous animals roaming the jungle). Visitors are not allowed to challenge the world that Disney presents; instead, they have to act like duped consumers in Disneyland.

In both Stitch Encounter and Adventureland, incongruity indexes the subordinated status of mainland Chinese visitors. Preening himself in front of the audience, Stitch manages to normalize his own imaginative behaviour and language while marginalizing the 'earthly' Chinese audience. Similar to Stitch, the boat captain introduces to passengers the 'world' beyond home. In both instances, the mainland Chinese visitors are reminded of their 'inferior status' for not being able to comprehend the future and the exotic world outside the PRC – hence reinforcing their yearning for something global.

The Festival of Lion King: The deliberately unexplained cultural logic

While having lunch at Disneyland, Lee overheard an older couple telling a staff member they wished there were tourist guide as they did not know what to do in Disneyland. Pointing at the map, the staff encouraged them to go to live shows; the older couple then left for the Golden Mickey Show. Unlike Stitch Encounter and Adventureland, most visitors have been to live shows, whether it is a community variety show, Cantonese opera or a musical. Even though the content may be foreign to aspiring global consumers, the format is not.

The Golden Mickey Show is well received because the story is universal – different Disney characters receiving awards. The Chinese hostess of the show uses Cantonese to introduce the characters while English and simplified Chinese translations are projected on both sides of the stage. The show consists of various performances, such as Mickey and Minnie Mouse dancing; songs from various Disney movies such as *Beauty and the Beast* and *The Hunchback of Notre Dame*; and acrobatics from Tarzan. Other than the hostess, none of the performing cast is Asian. To audience members who are not familiar with Disney films, the characters may be meaningless and the songs may not resonate, yet they can still enjoy the live performance. Like the princesses in the hotel lobby, the Caucasian performers reaffirm to the audience that they are having a global experience.

In contrast to the rather universal appeal of the Golden Mickey Show, the musical show Festival of Lion King is puzzling to those who have not watched *The Lion King*.

The music show does not follow a chronological event of *The Lion King* story; instead, the show begins with the finale of the original story, in which all animals in the African forest are having a triumphant celebration on the green Pride Lands with the grown-up Lion King Simba and Queen Nala both standing on the top of Pride Rocks to receive the applause of their people. The first scene ends with Rafiki (a wise old mandrill) presenting Simba and Nala's newborn cub, which symbolizes 'The Circle of Life'. Then the show returns to the scenes of the old days when Simba was still a cub misguided by his uncle that he was responsible for the death of his own father, Mufasa. There follows Simba's long survival journey with his friends Timon and Pumbaa, a warthog duo who adopt and raise the cub under their worry-free philosophy, Hakuna Matata.

The musical is conducted in English; an African American singer opens the stage pronouncing the authenticity of the show. Unlike the Golden Mickey Show, there is no Chinese translation. Occasionally the characters insert a scene summary in Cantonese or Putonghua, but no word-for-word translation is provided. Disney makes no attempt to explain the plot and the characters to the diverse audience.

The musical is unevenly received. The global consumers who have an understanding of the original *Lion King* story find the show easy to follow even though the story presented in the musical uses a different chronological order. As they do not need to spend extra energy to read the story, the global consumers can enjoy the pleasure, laughter and tears generated by the show. They can marvel at the imaginative adaptation of the story from screen to stage. However, the musical is puzzling to the mainland Chinese audience. First, owing to the language barrier, they are unable to understand the English lyrics and dialogues. Having not seen the Disney film, the audience, regardless of their competency in English, do not possess the required cultural resources to reorganize the events in the story, to recognize the characters' names, and to understand the jargon (e.g. Hakuna Matata) and the latent meanings of the story. Many mainland Chinese audiences at most could tell the narrative was a legend of a lion king. Using their cultural resources of Confucianism and Chinese ethics, they might only grasp the notion of championship of justice over wickedness, good over evil and harmony over disparity. When Fung asked a mainland Chinese audience member what he thought of the show, he said: 'I don't understand [the plot and story]. It is a game [and] a show. It doesn't matter. But I feel good about it. Great! Great!' (informal discussion on 7 July 2008). Many mainland Chinese audience members, however, left the venue before the lights were switched on. Yet this minimal understanding of the plot would justify their purchase of *Lion King* merchandise, fulfilling their role of the global consumers.

What is intriguing is that, even though few mainland Chinese tourists fully understand the story of the *Lion King*, they are highly satisfied with the performance. There are two possible explanations. First, children are usually excited about the show – no wonder Disney used a happy girl child's face to advertise Disneyland on mainland Chinese television. Second, without a conscious effort to appreciate the story, the audience values the exposure to a fresh, creative and professional musical. In other words, the new cultural form itself is more powerful than the content.

The Many Adventures of Winnie the Pooh in Fantasyland is another example of how aspiring global consumers may enjoy the form over the content. The ride takes the audience through A.A. Milne's story from one scene to another. The narration is strictly in English and there is no explanation of the story before the passengers go on the ride. To visitors who are not exposed to Winnie the Pooh and the story, the ride is nothing but a dazzling display of technology: one may wonder how the artificial wind is produced or how the doors automatically open and close. But an experience of technology itself is a step towards becoming a global consumer.

The globality for Chinese audiences

All Disneylands have to cater to both local and international visitors, and Hong Kong Disneyland is no exception. However, the park in Hong Kong also has to cater to a third group of visitors: mainland Chinese. Unlike both local and international visitors, mainland Chinese visitors are not yet global consumers: they do not possess adequate cultural resources to understand and enjoy Disneyland. Nonetheless, the amusement park has become a must-visit site in Hong Kong.

We posed the following questions: what is local culture? Is the boundary between local and global culture fixed? As a former British colony, Western education and culture have been imposed on Hong Kong people. As a free market and a global city, Hong Kong people are used to the flow of people, goods, cultures, media and technology. Like children in developed countries, most Hong Kong Chinese grow up with Disney characters as much as Japanese cartoons. Although Hong Kong children may not understand Disney in the same way as American children do, Disney culture is integrated seamlessly with their own. Adults who grew up before Hong Kong's economy took off in the 1970s are familiar with Disney characters but may not harbour a similar affection as the younger generation does. They see Disneyland as another entertainment spot, just like the local Ocean Park, shopping malls or countryside parks. Many adults visit Disneyland after being persuaded by friends and children. Their passion for the Disney brand is low but somehow they feel compelled to pay a visit under the influence of the media. The older generation of Hong Kong Chinese (like Lee's mother) is marginally familiar with Disney characters; they may recognize Mickey Mouse and Winnie the Pooh, but they do not know the stories behind them. Yet they may still pay a visit because of curiosity, or because their children take them there to show filial piety. To the local visitors, Disney's values and products pitch in with the hype-loving, popular culture-centric, commercialized values and lifestyles of Hong Kong Chinese. Hence, from Disney's perspective, there is no need to localize the park.

At the beginning, Hong Kong Disneyland received criticism – mainly from an economic standpoint – from the press and locals: the locals sneered at the small scale of Hong Kong Disneyland; and the media questioned whether the partnership with the Disney Company was a sound one for Hong Kong people. But Chinese audiences' perceptions of Hong Kong Disneyland are not negative. Unlike Hong Kong Chinese, mainland Chinese's contact with foreign culture is restricted by the PRC because of media control, importation of foreign culture and restrictions on travelling overseas. Although China has opened its doors to foreign investment, it has only partially opened its doors to transnational media companies (Fung 2008). Coincidentally or not, as a way to boost Hong Kong's economy – and also as a political means to maintain Hong Kong's economy prosperity – the PRC has loosened regulations governing travel to Hong Kong. Since July 2004, mainland Chinese have been able to come to this Special Administrative Region through the new Individual Visit Scheme that is applicable to residents of 30 cities in China. For these Chinese visitors, travelling to Hong Kong is like seeing the world on Chinese soil. Going to Hong Kong Disneyland allows mainland Chinese to have a global experience without going overseas. This is an easier way to get this global taste than applying for a visa to enter the United States.

Disney's strategy is to deliberately construct a wide cultural gap between the American dream and the veracities of mainland Chinese by privileging the former and disparaging the latter. Mainland Chinese would like to experience the 'real' Disney tradition – the one that has been constructed since 1928 when Walt Disney launched the beloved Mickey Mouse character. Like a lot of countries that experience rapid economic growth, such as the United States at the turn of the twentieth century and Japan after the Second World War, China is searching for its own modern cultural expression; the first step, unfortunately, is to copy from other cultures. Pirated versions of Disney characters are broadcast in different mainland channels; the popular, state-owned Shijinshan Entertainment Park simply unapologetically 'steals' Mickey Mouse, Tigre and the Seven Dwarfs without seeking permission from Disney. Hong Kong Disneyland is hence a temporary escapade of the malevolent culture. Mainland visitors are prepared for a cultural shock that is not available in mainland China. To foreground the difference between Disney's global dream and mainland Chinese visitors, Hong Kong Disneyland constantly reminds visitors that what they are consuming is uniquely American. Therefore, the closer Hong Kong Disneyland is to the 'original' parks in the United States, the more global the Chinese visitors feel. The Disney Company understands that the park exposes mainland Chinese visitors to a sense of being global, which is superior to being nationalist as promoted by the PRC.

To some mainland Chinese visitors, going to Hong Kong Disneyland is an experience of global culture, but not necessarily an enjoyment. To some, the primary purpose is to purchase Disney merchandise as souvenirs and gifts. Many others are uninterested visitors with a 'go and see' mentality; they do not intend to spend an entire day at the park. Many go to the park because it is included in the itinerary of a cheap packaged tour. Some Chinese business organizations even sponsor employees to visit Disneyland for free or at a very low price. There are also those who work for companies that give away a Hong Kong trip – and a sojourn at Disneyland as part of it – as an 'open secret bonus'. Unlike what Disney expected, mainland Chinese visitors only stop by Disneyland for a few hours. They have no anticipation about the rides and games, yet nor are they reluctant about the visit. Local critic Lo (2005) says that Hong Kong has become a theme park to mainland Chinese visitors: the global city itself serves as a spectacle, an entertainment and a difference. Hong Kong Disneyland is only one stop out of many.

Conclusion: Neo-globality over nationalism

Through an ethnographic study of Hong Kong Disneyland, we found that, unlike other transnational corporations, Disney deliberately 'localizes' by keeping its amusement park intact in Hong Kong. Hong Kong Disneyland has presented an American fantasy and imagination to the audiences, not a diluted or a modified version. Even though mainland Chinese visitors have little familiarity with Disney characters, the park assumes that the visitors fully understand the plot and legends of Disney animations – or that if they do not, one day they will. There is no apparent localization strategy; at most, it is only a segregation of Hong Kong, mainland Chinese and English-speaking visitors.

We asked at the beginning of the article why Disney refuses to localize the park and why local consumers are willing to subject themselves to homogenization. Current literature on localization assumes there is a well-defined boundary between local and global culture, and that local cultures are more desirable to the local audience. However, what if local cultures are not cultivated and fostered by the people but by the state, as in the case of China? In the case of Hong Kong Disneyland, the company's strategy is to amplify the uniqueness of Disney culture. While cultural critics may see Disney as a hegemonic force to impose its values on its audience, mainland Chinese visitors may view it as an alternative to a state-controlled and manufactured culture. Some audience consumers may prefer Disney's intact global culture to the nationalistic ideology. They may enjoy the entrapment into the global culture as a gesture against the disenfranchisement of Chinese politics, political participation and personal freedom. The audiences might not find Mickey Mouse particularly adorable, nor do they emotionally echo with the bubbling gaieties of the Disney Kingdom. However, what they experience is a *difference* – a free, individualist American atmosphere that sharply contrasts with the repressive state in which they reside. Unintended by Disney, the Hong Kong Disneyland may serve as an alternative, creative counterforce for mainland Chinese visitors.

Note

1. Highlights. Hong Kong Disneyland: An asset for the future (http://www.info.gov.hk/disneyland/eng.htm). Background. Hong Kong Disneyland: An asset for the future (http://www.info.gov.hk/disneyland/bi-e.htm).

References

Chan, J.M. 2002. Disneyfying and globalizing the Chinese legend Mulan: A study of transculturation. In *In search of boundaries: Communication, nation-states and cultural identities*, ed. J.M. Chan and B.T. McIntyre, 225–48. Westport, CT: Ablex.

Chan, J.M., and E. Ma. 2002. Transculturating modernity: A reinterpretation of cultural globalization. In *In search of boundaries: Communication, nation-states and cultural identities*, ed. J.M. Chan and B.T. McIntyre, 3–18. Westport, CT: Ablex.

Disney is optimistic about Hong Kong park. 2003. *Wall Street Journal*, 22 October: B2L.

Dorfman, A., and A. Mattelart. [1971] 1991. *How to read Donald Duck: Imperialist ideology in the Disney comic*. Trans. D. Kunzle, New York: International General.

Fowler, G.A. 2006. Disney fires a broadside at pirates. *Wall Street Journal*, 31 May: B3.

———. 2008. Main Street, HK: Disney localizes Mickey to boost its Hong Kong theme park. *Wall Street Journal*, 23 January: B1.

Friedman, J. 1995. Global system, globalization and the parameters of modernity. In *Global modernities*, ed. M. Featherstone, S. Lash, and R. Robertson, 69–90. London: Sage.

Fung, A. 2008. *Global capital, local culture: Transnational media corporations in China*. New York: Peter Lang.

Gerhson, R.A. 1997. *The transnational media corporation*. Mahwah, NJ: Lawrence Erlbaum.

Hilsenrath, J. E., and Z. Coleman. 1999. Disney Park deal may not wave a magic wand over Hong Kong. *Wall Street Journal*, 4 November: A26.

Lo, K. 2005. *Chinese face/off: The transnational popular culture of Hong Kong*. Urbana: University of Illinois Press.

Lull, J. 1995. *Media, communication, culture: A global approach*. New York: Columbia University Press.

McPhail, T.L. 2002. *Global communication: Theories, stakeholders, and trends*. Boston: Allyn & Bacon.

Morley, D., and K. Robins. 1995. *Spaces of identity: Global media, electronic landscapes and cultural boundaries*. London: Routledge.

Orwall, B. 1999. Walt Disney set to unveil park in Hong Kong. *Wall Street Journal*, 2 November: B11.

Pieterse, J.N. 1995. Globalization as hybridization. In *Global modernities*, ed. M. Featherstone, S. Lash, and R. Robertson, 45–68. London: Sage.

Reis, R. 2001. Brazil: Love it and hate it: Brazilians' ambiguous relationship with Disney. In *Dazzled by Disney? The Global Disney Audiences Project*, ed. J. Wasko, M. Phillips, and E. Meehan, 88–101. London: Leicester University Press.

Ritzer, G. 2008. *The McDonaldization of society*. 5th rev. ed. Thousand Oaks, CA: Pine Forge Press.

Sassen, S. 2001. Spatialities and temporalities of the global: Elements for a theorization. In *Globalization*, ed. A. Appadurai, 260–78. Durham, NC: Duke University Press.

Slater, J. 1999. Aieeyaaa! A mouse. *Far East Economic Review* 162, no. 45: 50–1.

Thussu, D. 2000. *International communication: Continuity and change*. London: Arnold.

Wasko, J. 2001. *Understanding Disney: The manufacture of fantasy*. Cambridge: Polity.

Wasko, J., M. Phillips and E.R. Meehan, eds. 2001. *Dazzled by Disney? The Global Disney Audiences Project*. London: Leicester University Press.

Winseck, D., and M. Pike. 2007. *Communication and empire: Media, markets, and global communication 1860–1930*. Durham, NC: Duke University Press.

Yoshimi, S. 2001. Japan: America in Japan/Japan. In *Dazzled by Disney? The Global Disney Audiences Project*, ed. J. Wasko, M. Phillips, and E. Meehan, 161–81. London: Leicester University Press.

Architecture on the move: Urban and architectural design in Inner Mongolia

Bert de Muynck

It's funny ... The desert's a big place, but nothing really ever gets lost there.
The English Patient, 1996

Introduction

Ordos, also known as Erdos, is located 80 minutes' flying time west of Beijing. With Baotou (100 kilometres directly north) and Hohhot (200 kilometres northeast), it completes a metropolitan region referred to as the Inner Mongolian Golden Triangle. The city sits at the top of the Ordos Basin, an area known as 'China's twenty-first century energy bank'. Today the city of Ordos has entered a stage of large-scale development. In the slipstream of a large number of energy projects, Ordos is building a city from scratch in the Kangbashi New Area. In 2008, as part of that development, 100 international architects were each invited to design a 1000 square metre villa, including a swimming pool, in an area designated as the 'Ordos Cultural Creative Industry Park'. Flying 100 architects from 27 different countries to Ordos twice in less than three months was already a unique undertaking. The word 'unprecedented' is too frequently used to characterize China's urban development, a blind and uncritical epithet for a country in the throes of change. But in Ordos, the 'ORDOS100' project was beginning to suggest that this word was meaningless. In January 2008, a first group of 28 architects was invited for a project introduction and site visit. They came back in April to present their proposals. Another 69 architects joined them during the April meeting, and were invited for a site visit and to see the first proposals. In June, 72 practices – three more having arrived in May – came to Ordos to present their proposals. In less than half a year, 100 villas by many of the world's young and promising architects had been designed and presented in a city in China few had ever heard of.

ORDOS100 is the brainchild of the Chinese tycoon Cai Jiang, who made his fortune in milk and coal and who is a passionate patron of architecture and the arts. He acts as the client in this project, which happens through the investment of two companies that he runs, namely Jiang Yuan Cultural & Creative Industrial Development Ltd and Jiang Yuan Water Engineering Ltd. During the January meeting, he explained the commission he gave to the architects as follows: 'We give the architects the freedom to design whatever they want, so they can put all their ideas into their work. If there happens to be a difference between the designs and the Chinese regulations, we will do something to make the balance and try our best to make it better.' The second player in ORDOS100 is the iconoclastic Chinese artist and architect Ai Weiwei, who acts as the curator of the project. In January he expressed his hope that ORDOS100 would contribute to the development of Chinese architecture through an exchange of knowledge. He told me that 'this project focuses both on China and the world. For me the important question is how to bring worldwide contemporary architectural knowledge in touch with Chinese practice. This is an important part of this project. To me it is really about action and the understanding of today's culture.' And last but not least, there is the context. Ordos is located in the Inner Mongolian desert, making it very cold (as low as $-40°C$) in winter and hot (up to $40°C$) in summer. For the majority of the architects, working in China means adapting to a different building process while acknowledging that their involvement ends with the design phase, as the construction drawings and the execution are the responsibility of their Chinese partners.

Context

About one hour's drive from the construction site of the ORDOS100 project is the Genghis Khan Mausoleum, a tourist spot that serves to remind us that this is the territory from where Genghis Khan sprang to conquer a large part of the known world. Today, the roles are reversed. Through the conquering of its own territory, the local government marks its ambition to claim a position in the world. The exploitation of copious resources – the area sits on top of the world's seventh largest coalfield (and China's largest) as well as on an enormous gas field – is the driving force behind the quantum leap in urban development that China's Inner Mongolia Autonomous Region will make over the next decade. Today, China is investing heavily in its energy development and the central government has listed optimizing the structure of China's energy resources as 'the priority among priorities'. The China.org website explains the situation as follows:

> Currently coal and oil products from the Ordos Basin have played an important role in China's economic development. The basin is now providing 4.75 billion cubic metres of natural gas to 15 large and medium-sized cities including Beijing and Tianjin. Natural gas will also be pumped to cities along the eastern coast such as Shanghai and Nanjing starting this October. (Xinhua News Agency 2003)

Today, the first results of this economic development can be seen. The rise of per capita GDP last year surpassed that of Beijing. Having the second highest per capita income in China and an annual economic growth rate of 40%, Ordos is capitalizing on its resources, investing in real estate construction, and creating a new middle class that needs better living conditions and a new generation of billionaires that wants outstanding buildings. One sign of this development is making the area accessible. The opening of Ordos Airport towards the end of 2007 was a first step to attract business and tourism. Although modest in size, Ordos Airport is China's first privately run airport with a total investment of RMB200 million featuring a 2400 metre runway and terminals covering 3500–5000 square metres.

The 100 architects were selected by Swiss architects Herzog & de Meuron. The invitees were not only asked to design in total freedom, but have clearly been identified as the creative messengers who will draw worldwide attention to this place. At least that's what the local government, the client and the curator want them to do. And given the media attention so far, they appear to be succeeding.[1] But in this case, it's not only marketing. The strategy of inviting internationally recognized architects to lend lustre to high-profile developments is common practice all over the world today. One of the many recent examples is Next Gene 20, a collection of 20 villas in Taiwan, designed by MVRDV, Kengo Kuma and Julien De Smedt, and the like.

But Ordos100 differs from similar projects in more than one respect. First, ORDOS100 is part of a whole new urban development of which a Creative District was part of the program to construct a new city from the start. Second, it's actually going to be built, and very quickly at that. According to planning, construction of the 100 villas will start after the Mongolian winter of 2008 and will be finished by the end of 2009. And third, Herzog & de Meuron's selection of the participating architects, following their design of the National Stadium (also known as the Birds' Nest), represents a new collaboration between Herzog & de Meuron and Ai Weiwei. In an interview some years ago, the Swiss architects explained their reason for working with the prominent Chinese artist-architect as follows:

> Weiwei is someone who tests our ideas. [Jacques Herzog] We have lengthy talks with him about how things work in China today. You cannot just walk into China and do what you have always done. We like to learn from other places, and China is the oldest civilization on the planet. With Ai Weiwei, we find contemporary lines of energy from that tradition. (Pearman 2004)

For ORDOS100, Herzog & de Meuron have avoided inviting *starchitects* and opted for young, up-and-coming architects instead. For the majority of these architects, working in China means adapting to a different building process from that to which they are accustomed.

The bulk of China's urban development happens through stitching urban development plans to existing cities. Ordos represents a rather unique case. In 2001, it became a Prefecture Level City (PLC), making it possible to have its own People's Congress, and granting the local government more power and ability to oversee city development. In March 2005, the *Erdos Urban Region Development Strategy – A Report to the Municipal People's Government of Erdos* was published. This report for 'The World Bank (EASUR) and The Cities Alliance' was prepared by Chreod Ltd (www.chreod.com), a company advertising itself on its website as 'an independent consulting firm providing market research, investment risk analysis, development planning, public policy, and information management services to corporations, governments, and international financial institutions active in China'. The report had the following objective:

> The scope of the Erdos Urban Region Development Strategy was defined during consultations with the Erdos CDS Leading Group and with stakeholders. Central to the strategy are the recommendations that will support the preparatory work for the Eleventh Five Year Plan (2006–2010). The City Development Strategy (CDS) is viewed as an important input to the long-term development plans for Erdos, with a timeframe to 2020.

Moreover, the report clearly states the ambition for the Erdos Urban Area to be enlarged to 42 square kilometres to support a population of 355,000 by 2020. It also states that the urban economic constellation will be 5:50:45 among the three sectors (primary, secondary and tertiary) and that the newly built district should promote Mongolian-style eco-tourism. The report even hints at the type of architecture this new city should develop: 'In fact, new buildings are to reflect elements of the Mongolian nationality.'

Growth of a region

On a half-hour drive from the new airport, through largely empty desert territory, one starts noticing construction cranes filling the horizon. Here, on a land area of 155 square kilometres, the local government is building a city from scratch with a planned population of 200,000 by the year 2020.[2] At the same time, national highways are being constructed and a railway is planned to run through the south of this area. Ambitions run high. Yang Hongyan, the vice-mayor of Ordos, explained to me during the inauguration ceremony of the ORDOS100 project that 'with this new development we aim for excellence and exception'. I interviewed her in the Holiday Inn, where in 2008, in January (over three days), April (five days) and June (five days), 100 international architects came to attend conferences, engage in eco-tourism, present their projects, visit the Genghis Khan Mausoleum and criticize, along with colleagues, their own and others' design proposals. For many of them it was their first time in China, and for all except one the first time in Ordos. The vice-mayor continued:

> Three years ago (2005) few people would have been able to identify Ordos on the map. We built up a team, and the mayor proposed the concept that Ordos should reach out to both China and the world. Due to the economic development of this region we have an advantage and are aiming to attract companies in the top world 500. For planning the new urban district, we invited international architects to come in first, not just domestic. The development of Ordos is happening through leaps and bounds; we have ambitions to become one of the most vibrant cities in the west of China.

This ambition to develop a new model of a city has an advantage. As Ordos is a latecomer in its urban development, it can benefit from the mistakes and lessons learned by the Chinese cities that have developed, expanded and reinvented themselves during the past two decades. In that development, Ordos doesn't want to go unnoticed, but envisages that it will be both a test case and a role model, as the vice mayor explained:

> In the 1980s we looked at Shenzhen as a model for urban development, in the 1990s we looked at Shanghai and Pudong and it is my hope that in the coming 20 years when people look for a new model they will look at the development of Ordos.

In the first half of 2008 I visited Ordos three times and took several tours through this city under construction. From the aforementioned report it is clear that the local government is serious about implementing the second of 'two important government plans for the Erdos Urban Area':

> Second, the new Erdos Municipal Government (of the entire PLC) proposes to develop the Qing Chun Development Zone or New Area (QNA), also known as the Kangbashi New Area. The QNA is to be located 27 kilometres away from the current city and will have a targeted future population of approximately 150,000. This new town idea rests on a number of presumptions: relocation for water access, escalating population densities in the Erdos Urban Area, and the cost-effectiveness of developing new land in Qing Chun, as well as recognizing that the municipal government needs a new administrative centre due to existing ambiguities surrounding revenue streams and government restructuring when Erdos attained PLC status. This development proposal has been approved by the State Council. The total area of the new development is to be 200 square kilometres, with a core urban area covering 31 square kilometres. (*Erdos Urban Region Development Strategy*)

That the invited architects are brought together at the Holiday Inn during their stay in Ordos is no coincidence if we read the first of 'two important government plans for the Erdos Urban Area':

> The first, an initiative of the Dongshen District Government, is to develop the Tend Xi Re Area (TA). The TA is located west of the railroad which acts as a border to the existing urban area. The total planned area is 15 square kilometres. Currently, 7 square kilometres are serviced with roads and electricity. A five-star hotel has been built and is now operating as a Holiday Inn. The government has financed the construction of this facility. (*Erdos Urban Region Development Strategy*)

Constructing creativity

One important feature of the Ordos urban development is the integration into its urban program of the cultural and creative industries. In 2005, the State Council (China) released its '11th Five-year Plan' and put the creative industries on the formal agenda. A couple of months later, in his address to the 17th National Congress, President Hu Jintao stressed developing the cultural industries as a means to enhance culture as part of the soft power of China. These – sometimes referred to as 'cultural', at other times as 'creative' – programs fill both a void in the existing programs of Chinese cities and attract further real-estate investment like shopping malls, retail and mixed-use developments, targeted to a growing mass of Chinese consumers.

The Chinese city is a space of conflict, confusion, crowds, culture and construction sites. Despite its exceptional development and growth, the Chinese city has the same programmatic characteristics as the usual city: housing, offices, parks and roads. Scale and the pace of construction distinguish the Chinese city from the usual city. In Beijing, a group of artists accidentally discovered a new urban program, a way of living between art and economy. A couple of years ago, they transformed an old industrial factory into an experimental laboratory. Eager to capitalize on their creativity, they failed to foresee how their act of innovation would destroy the source of this creativity: the place itself. The art factory soon turned into an art market. Art was produced in another part of town, but still consumed in the factory. In the end, it was all about place-making, branding and imposing international policies upon a local context. Today, it doesn't matter what is on display, as it is about a brand – in the Beijing case 798 – and the creative industries. Once that formula was understood, tested and controlled, it served as a model radiating from Beijing outwards. This led to a formulation of the future of the Chinese city, a city where creative business districts (CBDs) and special creative zones (SCZs) are an indispensable part of urban planning. These areas offer everything between creativity and consumption, folk cultures and foreign intrigue, coffee and cultural critique.

The Ordos Cultural Creative Industry Park has a total planned area of 197 hectares, and a total construction area of 1.46 million square metres. Next to the focus on creativity, there is also a plan for SOHO1 and SOHO2. With a total investment of RMB4.5 billion, the construction plan will happen in three phases. The first construction on the site started in July 2006 with the Ordos Art Museum, which opened in August 2007. ORDOS100 is part of the first phase of this urban development. Phase two is expected to be completed in 2011, and phase three in 2012. The idea of bringing in 100 architects is to create diversity in quality and creativity. However, this seems to be at odds with the general image of the work of foreign architects in China. Mostly they receive commissions and requests to design larger projects than they would ever be able to build in their home countries. Designing a villa at an almost record pace in conditions with which they are unfamiliar will undoubtedly make the area distinct in character. The outcome of this project is a high-end residential setting, which some see as akin to a World Expo, while to others it looks more like Beverly Hills or an Architectural Zoo where international ideas on housing can be tested under Chinese conditions of speed, quality of construction, labour skills, a creative industries context and return on investment, as each of the 100 villas is expected to sell for $1.5 million. At the same time, this development is in tune with the demand of the market for the wealthy who look for culture, second, third or fourth residences and uniqueness in their lifestyle.

First impressions

In January 2008 a site visit with the participating architects was arranged. At the time there was a largely empty site where one could notice the unique desert topography and climate characteristics, as it was − 20°C. It was stressed to the architects that they needed to incorporate these severe climate conditions into their designs. The architects knew nothing about the future inhabitants of their villas, except that they were billionaires. Two buildings bordered the site: the Ordos Art Museum designed by the Chinese architecture firm DnA and the Artist Studios by Ai Weiwei/Fake Design, the last one being an exact replica of the artist studios that Ai Weiwei built in the upcoming Chaochangdi art district several minutes' drive from Beijing's notorious 798 art district. The Ordos Art Museum features a 2700 square metre space for an exhibition and research program. For some architects, it is hard to imagine the transformation of this barren site. 'I need all the imagination in the world', Dutch architect Kamiel Klaasse from NL Architects tells me. 'Is this going to be a rough place or a sophisticated paradise? It's hard to imagine what our client sees beyond the horizon.' The high density of the master plan confuses some architects as well. Mexican architect Derek Dellekamp said:

> Considering the fact that there is no lack of space here, I expected the villas to be more free standing. The openness here is a treasure. Why not use it? According to the master plan, this area will be very urban. I have to make a mental switch for that.

The fact that there are 100 architects involved raises some concerns as well. 'I can't deny that our client has a fantastic vision', says Indian architect Gurjit Singh Matharoo. 'But there could be big chaos in the end. There are so many minds on such a small plot of land. It might not generate a beautiful place, but maybe this project is about something else, about the exchange of architectural ideas, for instance.'

Three months later, when we meet again in the Holiday Inn hotel lobby, Matharoo proves to be right. The atmosphere is both relaxed and tense. Some architects (those who made a design for the first phase) are here for the second time; others (those who worked on the second phase) are here for the first time and curious to understand what is happening. In the corridors, the proposals of phase one are displayed for discussion. During the arguments, one hears English, French, Spanish, German, Dutch, Hebrew, Swedish, Chinese and some other languages. It becomes clear that ORDOS100 is an attempt to build a Babel for billionaires and Ai Weiwei is its curator. After the dinner on the first night in April, the architects from phase one fill a large mock-up of the building site with their models. A whole caboodle of architecture arises and different styles compete with each other: a villa with 100 rooms, a dwelling without distinction between inside and outside, a house with a green heart, a monolith, different boxes colliding together into one form, a house based on the idea of 'holistic materiality', a villa without a claim on its territory, a dwelling dug into a dune, a green mountain rising out of the desert. In June, a stack house, a knot house, a twice house, a house house, a mirror garden house, a big brother house, and a *hutong* on top of a union are some of the villa concepts added to the site.

The Parisian office Encore Heureux, along with G-Studio, designed a Gourbi Palace. The architects call their retro-futurist design a 'survival-utopia': a 1.4 metre brick wall protects the inside from the outside, shielding it from warmth, cold and wind. Their justification:

> The notion of context is absent as we didn't know what other architects would propose. For us the context is the sky, as we were sure nobody would touch that. So we made this central void, around which the house is organized.

Israeli-Palestinian architect Senan Abdelqader proposes a reinterpretation of the traditional Chinese villa. 'Globalization is all pervasive', he tells me. 'I think that the Chinese dwelling, like the Arab house, should be translated into a modern idiom. I don't believe in the vernacular anymore.' A series of voids and solids vaguely resembles a traditional house and creates a compact plan that maintains a low profile. The Mexican office Productora opts for a narcissistic house that brings the façade to the interior. The explanation: 'By slicing up the volume, we created an introverted house as each room looks at the back of the next segment, a closed brick wall. When you enter a house you normally lose sight of the façade, but here one is constantly confronted with it.' And these are only the architects of phase one.

As soon as the architects from phase two arrive at the meeting in April, the Babel-like confusion is complete. Some want to critically review the assignment and the master plan; others just want to give an answer to the brief. Turkish architect Han Tumertekin (Mimarlar) is an exponent of the latter. 'When I am confronted with such a big organization I presume the assignment is well considered', he says. 'Architects have become much more interested in issues outside architecture over the past few years, but some of them have forgotten how to carry out a brief well.' Chilean architect Alejandro Aravena (Elemental) is also supportive of the master plan. 'Luxury houses in such a dense context are unusual for wealthy people. I hope this could shift the tendency of rich people to try and isolate themselves from each other.' Korean architect Minsuk Cho (Mass Studies) sees another challenge: 'Normally a city is composed of layers of accumulated history. Here we have to come up with instant identity, character and all of that. How do you do that? I don't know yet.' But whatever the results, ORDOS100 is, from all perspectives, a unique project. It gives the architects involved a lot of freedom and asks them to push the limits of their creativity, while at the same time it forces them to come up with a strategy for the lack of control during the building process.

Some of the architects involved in phase one have taken a close look at Ai Weiwei's own work and his philosophy of keeping architecture simple. One of them is Senan Abdelqader. 'My plan is clear and uncomplicated', he says, 'so it should be very easy for Chinese workers to realize. That's my way of making sure that the result resembles my intentions.' What is to come out of it remains to be seen, but according to Ai Weiwei this uncertainty is exactly the point of the whole project. When I interviewed him in October 2007, he told me he would quit architecture. When I confront him with this statement in April 2008, he laughs: 'I am not on the train any more, I am working on the railroad now', he replies. Meaning he now shapes the conditions for architects to do their work. He quit architecture and is now working on the 'railroad', because he wants architects to meet, discuss and create without constraints. 'We provide possibilities', he tells me with conviction. 'That's what I am interested in. We need possibilities instead of conclusions and results.'

A city from scratch

How do you build a city from scratch? During the past year criticism has been voiced on the way China has been developing its cities, most notably the notorious critique by Qiu Baoxing, China's vice-minister of construction, that 'many cities have a similar construction style. It is like a thousand cities having the same appearance' (qtd in 'Rush to modernity' 2007). Ordos is hoping to make a difference, creating a unique urban setting by focusing both on the rich cultural legacy of the area as well as connecting it with the best of the West. Singaporean firm AGV is carrying out the planning of the new district. Its master plan is to organize along a concentric pattern and based on the idea that the new city centre is like a sun rising from a meadow and radiating from the centre to all sides. In the heart of the project lies a central axis, connecting the governmental administration area with the entertainment area and a financial district. The central square – Genghis Kahn Square – is 1.6 kilometres long and 400 metres wide. Around this square are currently four buildings under construction called the four cultural relics buildings: a library, museum, culture and art centre, and ethnic theatre. They have been designed by Beijing- and Shanghai-based firms. Here again the architecture holds references, albeit maybe a bit too explicitly, but certainly in line with the idea that 'new buildings are to reflect elements of the Mongolian nationality'; the design of the library evokes the image of the three most cherished ancient bibles for Mongolian people: the *Mongolian Mystery*, the *Mongolian History of Gold*, and the *Mongolian History of Origin*. Two of the texts were written by ancient residents of Ordos. The design of the ethnic theatre has the shape of the hat that traditional Mongolian people wear. Construction started in 2006 and this new administrative and entertainment centre of the city was expected to be finished in 2008. The new city is divided into two sections by large avenues and two green belts, one 100 metres and the other 400 metres wide. Further from the centre, new high-tech areas are being planned, along with an area for the emerging automobile industry.

The architecture of adaptation

The invited architects all operate in vastly different conditions; most work exclusively in their country of origin. Unsurprisingly, these architects travel with a set of ideas and design skills that now have to be incorporated within the Chinese city. To some, this import–export aspect of architecture doesn't pose a problem, as Mexican architect Julio Amezcua (AT103) explains to me:

> We find a lot of similarities here to Mexico: the way we interact, build and communicate in terms of 'Yes, yes we are going to do it like that.' When you check the final result it always looks different, but at least they put the risk to do it, they didn't stop it. Also, in our culture a lot of the people start building based on the rendering, which can really be a problem.

Chilean architect Alejandro Aravena talks about the same loss of control when he explains the way his villa should be constructed:

> A key issue in this project is how to manage distance. In my project, brick is the common language that shortens the distance and guarantees a quality of the design. I found a brick on the site and it is fantastic to see how easily it breaks, showing red on the outside, black on the inside. The breaking is very rough, construction workers can do it and place it on the outside. So it has this rough quality, independent if mistakes are made or not.

When asked to respond to a possible critique of ORDOS100, which questions the creativity invested in building luxury houses for billionaires in a country that is still largely developing, Serbian architect Milica Topalovic gives me the following response:

Obviously this was the first question I asked myself when accepting the invitation. From a European and American perspective this type of work would be extremely debatable, heavily criticized and probably not possible at all. Imagine to develop such a project in The Netherlands or Switzerland where any development is constrained by a much more intense public debate. There each project is the result of a kind of negotiation of public values of which nobody knows anymore what they are. Somehow in Europe the post-socialist democratic situation has led to a form of nostalgia featuring a conscious presence which keeps things more moderate. Here, I was curious to see how much these parameters are shifted and what direction they take.

On the level of architectural creativity implemented in the design, it is clear that for every participating architect there is another type of architecture, not only by way of giving names to the projects but also in the way the architects explain their concept, analysis and proposal during the presentations. Spanish architect Tony Montes Boada (F541, with offices both in Barcelona and Boston) sees the following distinction:

> What I find interesting is that although we all have the same commission and constraints, all projects look radically different. Besides these basic constraints, the notion of the local context is not so obviously visible in the proposals. I find it curious to be able to easily recognize the place where the teams come from, where they have been studying. For example, the projects from architects studying in the Ivy League architecture schools, the Swiss or Japanese projects. For me the cultural context relates to the place where the teams come from.

Increasingly, architecture is becoming a profession of managing, creating and controlling reality from a distance. Whatever creativity the architect gains, he or she soon has to relinquish in terms of control. ORDOS100 and its Cultural Creative Industry Park provide an interesting test case in the field of architecture to understand the impact of China's creative industries ambitions, as an urban and cultural model, as a real estate investment (Mr Cai Jiang), a curatorial practice (Ai Weiwei) and policy implementation (the local government of Ordos). ORDOS100 is branded and legitimated by the involvement of '100 International Architects' (note that they aren't called 'Foreign Architects'), positioning this project as an import–export experiment at the centre of the debate on the creative industries. It is a project that imports, exports, adapts and experiments with our understanding of exchange and development in the field of architecture, labour, culture, media and urbanism. There is certainly an element of creative roughness, if not brutality, in inviting 100 architects for this project. As it currently stands, this project is clearly an experiment in the production of a new architectural culture, irrespective of its shortcomings. As always, it is only through mistakes that we learn to make better cities and update urban models.

The House House, by American practice JohnstonMarkLee, deals with a doubling of the iconic house with the pitched roof, leading to a result Sharon Johnston describes as a 'primitive form that through a process of multiplying has the ability to absorb a big range of complexity and resolve this at the same time on different levels. And, it has an imaginary quality.' The proposal of the Portuguese practice SAMI Arquitectos is a house built up out of a single line, determining façades, floor plans and program distribution: 'We designed in a pure architectural way and saw it as an exercise with our imagination, background and ideas on how to do a house in China. What could be an interesting way to work with this? There is a thin border between architecture, art, and sculpture, as Ai Weiwei said in the first meeting.' For the Japanese practice Bow-Wow, there is another aspect to the commission besides, but influencing, the design. Faced with the intensity of human exchange in both gatherings, they refer to ORDOS100 as a 'social sculpture': 'We are interested in this aspect of the project. Participation is the most important, to become the material itself. Competition here is not so important. When we realised this we decided to create something calm and follow the restrictions of the masterplan.'

Conclusion

ORDOS100 will take up a unique position within international and Chinese architectural history. It will be a theme park of a particular kind, but also home to many. At the same time, it is too early to draw conclusions based on the three meetings, as for every architect there is a story attached to their villa in China – some realistic, others overly formalistic and some unrealistic. Some have raised concerns about the notion of sustainability or the fact that this is a gated community for billionaires. It seems that the first needs to be dealt with at the level of each individual villa, while the second is unavoidable given the Chinese context. The careful selection of internationally upcoming and locally established young architectural practices has created an exciting setting for experiment of all kinds: in cultural exchange, architectural discussions on sustainability, gated communities, architectural construction, the master plan mixed with the designers' desire to rethink the concept of the villa in a Chinese context which is characterized by a desire for innovation, alternatives, distance, local culture and context. For many of the participating architects, it is a great way to have almost carte blanche to express their desires and test them in the unique conditions of contemporary China. More than an 'architectural zoo', one gets the feeling of an 'architectural jewel box' – a series of precious villas placed in close proximity to each other in the wide open desert. Once all of them are built, our understanding of the impact, legacy and reality of ORDOS100 might change again.

The tremendous speed at which China has been urbanizing in recent decades has put to the test the idea of the city and has led to diverse visions of real estate investment. The unique opportunity to build a city from nothing in the desert of Inner Mongolia is the beginning of a new chapter of urban development in China. The introduction of international architects in a small part of this urban planning, which is an officially designated area for culture and creativity, is both a sign of intelligent city branding and will serve as a lever for further real estate investment in the area. With inhabitants who can afford a 1000 square metre villa, there is no doubt that other real estate investors will jump on to this train of establishing creative business districts, understanding that quality of design can make a quantitative financial difference.

Acknowledgements

In the context of ORDOS100, MovingCities has acted as embedded architects. While the term could be applied to many historical interactions between journalists and military personnel, it first came to be used in the media coverage of the 2003 invasion of Iraq. In order to find new ways of reporting on the state of the contemporary city, MovingCities drives on the philosophy of embedded architectural journalism, travelling to, reporting from and engaging with those places where architecture and cities are questioned, in the making or absent.

This paper draws from former analysis and writings developed around the subject that were published by Bert de Muynck/MovingCities in *MARK Magazine* (NL), *Perspective Magazine* (HK), *Urban China* (CN), *ArtForum* (CN) and the MovingCities website throughout 2008.

Notes

1. A few examples of the worldwide media attention generated through ORDOS100 include 'Xanadu 2.0' (*Urbane China* [CN], January 2008); 'Herzog Picks 100 Architects for Inner Mongolia Housing Project' (*Building Design* [UK], February 2008); 'DRDH to Design Mongolian Villa' (*The Architects' Journal* [UK], March 2008), 'Young Residential Architects Invade Mongolia' (*Architectural Record* [US], March 2008), 'Dawn of New Century: Ordos100' (*Urbane China* [CN], April 2008), 'In Inner Mongolia, Pushing Architecture's Outer Limits' (*New York Times* [US], May 2008) and 'Ordos 100' (*Icon* [UK], June 2008).

2. Depending on the source of the targets, figures regarding the size of the development and number of inhabitants for the new district vary. A general tendency is that the new district will have a size between 150 and 200 square kilometres and is intended for a population of between 150,000 and 200,000 inhabitants.

References

Erdos Urban Region Development Strategy – A report to the Municipal People's Government of Erdos. http://www.citiesalliance.org/cdsdb.nsf/Attachments/China+-+City-Regional+Development+Strategies+-+Report+Vol+2:+CDS+Erdos/$File/Vol+2_ED_Erdos_30Mar5.pdf

Muynck, B., de. 2008. I jumped on the wrong train: An interview with Ai Weiwei. *MARK Magazine* 12 (February–March).

Pearman, H. 2004. *Iconclasm rules: how Herzog and deMeuron work with conceptual artist Ai Weiwei on Beijing's new Olympic Stadium.* http://www.hughpearman.com/articles5/weiwei.htm (accessed 22 March 2009).

Rush to modernity devastating China's cultural heritage. *The Guardian,* 11 June.

Xinhua News Agency. 2003. Ordos basin undergoes large-scale development. 14 April. http://www.china.org.cn/archive/2003-04/14/content_1062131.htm

Great adaptations: China's creative clusters and the new social contract

Michael Keane

In 2007, in his keynote speech made on behalf of the 16th CPC Central Committee to the 17th National Congress of the Communist Party of China (CPC), President Hu Jintao stressed the need to enhance Chinese culture as the country's 'soft power':

> Culture has become a more and more important source of national cohesion and creativity and a factor of growing significance in the competition in overall national strength … [We must] enhance culture as part of the soft power of our country to better guarantee the people's basic cultural rights and interests. (Hu 2007)

Along with the appropriation of the idea of soft power, President Hu has also regularly stressed another key concept, that of 'autonomous innovation' (*zizhu chuangxin*). While ostensibly directed at the science and technology sectors, this nevertheless carries forward into the national debates about the need to create new ideas.

This paper looks at the emergence of hundreds of designated cultural creative clusters – art centres, animation bases, cultural zones, and incubators – most of which have mushroomed from disused urban industrial sites in the past three years. The cluster boom, which is predicated on the idea of the formatting of cultural production and consumption, has important implications for how we understand China going forward into the second decade of the twenty-first century. The cluster phenomenon has resulted in a substantive remaking of the social contract between officials, entrepreneurs, local residents, academics – and most significantly cultural producers.

In this paper I argue that the term cluster, taken from economic geography and grafted on to the cultural field, allows an almost seamless transition from socialist ideals of collectivism and cooperation (communes, town and village enterprises) into a capitalist growth engine. Artists, designers and creative entrepreneurs are navigating the socialist bus aided by a political road map called the Eleventh Five Year Plan. The clusters are containers into which local cultural characteristics are mixed. For China, the key idea is the template, so much so that there is a perennial search for the appropriate model. In the new era of culturally determined soft power the creative cluster is a symbol of artistic renewal. The clusters in turn enfold into zones, districts, parks, bases and spaces.

Along with these developments has come willingness on the part of propaganda officials to tolerate, even promote, political expression. The question that emerges here is: how did artists suddenly become a new productive force in the Chinese-century script? How did the creative 'super-sign' (Liu 2004) take root so suddenly – and how did this idea come ready-made from the United Kingdom, despite the sustained efforts of the commanding heights of US entertainment industries over two decades to break down the great propaganda wall?

The import–export trade in ideas

During its long history of contact with outside nations, China has excelled in taking in and adapting ideas. From the period of the Silk Road in the Han Dynasty (206 BC–AD 220) to the great openness of the Tang (AD 618–907) and Song Dynasties (AD 969–1279), China was a receiver of the best from the rest of the world. In the Yuan Dynasty, the Viennese traveller Marco Polo visited China and wrote exotic tales of the wonders of the East. Trade in ideas and artefacts increased gradually. By the Qing Dynasty (AD 1644–1911) a term had emerged to justify China's craving for Western knowledge – *xixue wei yong, zhongxue wei ti*: Western learning (technology) was good for its usefulness, but Chinese culture remained the core. Emperor Kangxi was suitably unimpressed: '[Even though some of the Western methods are different from our own, and may even be an improvement, there is little about them that is new. The principles of mathematics all derive from the *Book of Changes*, and the Western methods are Chinese in origin …' (Spence 1974, 74).

Foreign ideas nevertheless flooded into China as the imperial order imploded. The modern printing press allowed wide dissemination. The practice of adapting foreign ideas was put on hold as China pursued its experiment with Soviet Marxism, which was itself adapted into Marxist-Leninist Mao Zedong Thought. During its period of isolation from the West, beginning in 1949 and lasting three decades, China imported very few cultural ideas and formats, with the notable exception of Soviet socialist realism. Nor did it export its ideas to other nations, unless of course we count international propaganda services.

With the passing of Chairman Mao Zedong in 1976, China emerged from the chaotic Great Proletarian Cultural Revolution. A period of regime transition ensued during the next two years. By the end of this, the remnants of the Maoist revolutionary era – the so-called Gang of Four – were shown the back door. Soon after accepting the position of paramount leader, Deng Xiaoping moved to open the front door to Western ideas, at least those that were deemed complementary to the reform agenda. Ideas about development during the 1980s remained firmly linked to the key platform of modernization. Culture was still the superstructure, reflecting the base, which was about incremental economic reform, beginning in the countryside and moving to the cities.

As far as the much of the developed world is concerned, 1989 was an indictment of the true colours of the Chinese Communist Party. Openness was put aside as People's Liberation Army troops relentlessly put down the student democracy protestors in Tiananmen Square. During the early 1990s, China's propaganda leaders were regularly heard associating Western culture with sugar-coated bullets. People of China: beware the lure of Disney and Dallas. Meanwhile, ideas about development, the need to catch up in the fields of technology, and the beginnings of a preoccupation with scientific innovation were leading China closer to a new cultural frontier. As China's accession to the World Trade Organization (WTO) loomed, the core intellectual debate moved to the problem of how to preserve national culture from the 'wolves at the door', a term used to describe the capitalist entertainment complex, often represented by the Hollywood studios and epitomized by Rupert Murdoch's dogged determination to conquer China (Rosen 2002; Zhao 2002; Keane 2002).

Knowledges of the creative economy

China joined the WTO in December 2001 and a couple of years passed. The much-feared Western cultural invasion failed to materialize. Chinese culture remained strong and resilient. It was time to invest in cultural development. During this period the idea of the cultural industries was official recognized within the economic development agenda. It became a new way to think about developing international cultural trade, another idea that had moved on to the policy agenda in the period leading up to WTO accession. Concurrent with accession to the world's premier trading club, the focus turned from globalization to internationalization, from threat to opportunity. In the years following China's accession to the WTO, the term 'international' (*guoji*) became an obligatory description for promoting conferences and forums in which there were foreign speakers.

In the post-WTO period, however, a different soft invasion was heading east, this time not from the United States but from 'cool Britannia' via a Southeast Asian trade route. These 'knowledges of the creative economy' would have a deeper impact than the determined efforts of the Motion Picture Association of America to open China's cultural markets during the previous decade.

The 'creative industries' reached China in early 2004. By the end of the year this new industrial development idea, or, more appropriately, this package of ideas, had take a foothold in Shanghai policy circles. In the preceding years the creative industries had travelled gradually from their home base in the United Kingdom to Singapore and Hong Kong, with some degree of uptake in Australia and New Zealand. Kong et al. (2006) have written an interesting account of how these 'knowledge' ingredients were selectively adapted in East Asia. The term 'creative industries' had come from the United Kingdom with a mysterious and powerful invocation: 'those industries which have their origin in individual creativity, skill and talent and which have a potential for wealth and job creation through the generation and exploitation of intellectual property' (DCMS 2001, 4).

The new doctrine was led by scholar-consultants, many of whom spoke with British accents. International experts appeared offering advice about creativity indexes, creative classes, milieus, clusters, and networks. Conference organizers advertised for foreign speakers who could address the idea of creativity, often resulting in a mixed bag of confused generalizations from people who had no connection to the origins of the creative industries. Moreover, what was frequently lacking in the advice given was the specificity of Asian culture and politics. But advice was nevertheless received, absorbed, and adapted into the socialist stew that within two years would become China's 'cultural creative industries'.

In contrast to the more nationalistic cultural industries, the creative industries were championed by local and city governments. A short list of its benefits for cities included wealth creation, redesign of urban space, renewal of traditional resources, and clean service industries, providing more 'value-added', a cliché that came with the UK creative industries bible. The how-to-do-it guide to becoming creative was more than just about putting new wine in old policy bottles. Creativity was a green idea in a country where the 'Made in China' model had turned skies a brownish grey. Creativity was an unlimited resource, if it could just be tapped into. However, it also needed to be understood. What exactly were the creative industries?

While a discussion of 'Chinese creativity' is beyond the scope of this paper it is worth outlining its association with the theme of cultural adaptation. From the examples below we will see that a key to understanding creativity in China is the use of resources in an efficient way. Rather than making something new, the foundation of the Western Romantic tradition, Chinese creativity is about rearrangement according to circumstances, which may be political, social or economic. Such rearrangement, while always new in a certain sense, proceeds in patterns that are essentially recombinant. In China it is not so much originality that is sought but creativity that is appropriate to the context (Sigurðsson 2008). In the early twentieth century the great writer Lu Xun had drawn attention to this borrowed aspect of creativity. He advocated 'taking' (*nalai zhuyi*) as a bridge to develop China.

The adaptive potential of the creative industries was therefore 'there for the taking'. Whereas the cultural industries had captured the eye of Communist Party officials looking for a way to commercialize tradition, the creative industries came in a ready-made format. Moreover, it was an international idea, which added to its appeal. As opposed to the dank smell of museums, cheap souvenirs and the cultural relics of traditional culture, the creative industries format included incubators, lofts, digital content studios, and most of all, intellectual property. The creative industries format also included new theory – about wealth creation and increasing returns, rather than the old Marxist stages-of-development approach. If there was a way to leapfrog, it was on the back of this shiny idea. The creative industries bible had methodological revelations: how to understand the role of intellectual property, how to map the value of the various sectors – there were thirteen sectors – and how to construct creative clusters. This was utopian, a forward vision for cities languishing in the Asian economic crisis.

Inevitably, there was contention about which was the best 'industrial' format for a country with a 4000-year tradition: cultural or creative. The more Chinese socialist approach favoured cultural, with its evocation of 'the people', while the business approach advocated creative, for all of the reasons mentioned above (see Chang 2008). In the end the compromise 'cultural creative industries' (CCI) came into being, into which the myriad elements associated with culture and creativity were diced, mixed and enfolded. Even the letters of the various acronyms were adaptable. Adding 'international' to the mix produced the ICIA (International Creative Industries Alliance), the ICIC (International Creative Industries Conference), the ICCIC (International Cultural Creative Industries Conference), and the IACCIC (International Association of Cultural Creative Industry Clusters), not to mention that the 'CI' could also be used to refer to China, content, and copyright industries when the occasion suited. Indeed, as the logo of the first ICIC in China creatively proclaimed: 'I see, I see'.

Having envisioned CCI as the development template, the question still remained as to how it would adapt to Chinese conditions. The solution for China was to champion the idea of the 'creative cluster', banking on the existing success of industrial clusters in China's manufacturing economy, and typified by the low-cost Made in China brand (Keane 2007). Indeed, the term 'creative industry' was taken quite literally. Creativity had become an industry and as such it had industrial characteristics. But would the clustering of creative labour result in brand development or just large quantities of low-value cultural goods?

In order to better understand this dilemma, it is helpful to look briefly at the Chinese creative ecology. In short, this refers to China's movement up the value chain, from world factory to world studio, from OEM (outsourced production service) to ODM (original design), from factory to cluster (Keane 2007). Significantly, the direct association of creativity with the industrial economy in China overlooks many non-industrial manifestations of creativity. As Caves (2000) has demonstrated, these are atypical industries. Consumer demand is uncertain and success cannot simply be predicted; people are often not driven by the bottom line but 'for art's sake'. The question remained: how can creativity be exploited to produce competitive advantages? The obvious answer in China was to ignore the more idiosyncratic elements of creativity and opt for scale and agglomeration.

Clustering was quickly identified as a way of turning the intangible and the mysterious attributes of creativity into material forms (paintings, artefacts, sculptures) – in other words, 'things' that Chinese officials were familiar with. Cluster master plans circulated as factories were turned over to developers. State-owned enterprises, private business entrepreneurs, and university research centres moved into the creative field as a result of generous incentives from local governments (Keane 2007). Setting up a factory, calling it a cluster, and producing contracted products was an obvious business model. International animation companies, design firms and even movie production companies were willing to outsource to China. One example of such a 'creative factory' is Dafen, an artists' village in Shenzhen, south China, where approximately 8000 people work on replicating oil paintings for international markets. The artisans here have no art college training in the techniques of painting. On-the-job training provides the incentive. The work is produced for the most part in factory conditions. People work long hours replicating old masters as well as the kind of kitsch paintings found globally in cheap marts. Elsewhere, animation factories were competing to provide cheap outsourcing options for international companies, offering huge savings in labour costs. Indeed, between the low-cost factory at one end of the spectrum, and the creative cluster producing ideas at the other, there are many similarities.

The creative cluster

It is the creative cluster model that is of most interest in this paper. I believe it represents a pragmatic response to employing populations. Because of the relative newness of the idea of creativity, and the government's newfound support for artists, there is convergence between aspirations to broker new ideas and products and the fundamental problem of absorbing labour into new industrial sectors.

Clusters, from this perspective, are methods for assembling and managing creative labour.

As Ryan (1992) has pointed out, the legacy of organizing creative labour goes back to the nineteenth century in the United Kingdom and Europe. The irrational and often non-productive habits of artistic types needed to be managed and this led to a new class of business intermediaries, managers, agents and middlemen. Moreover, organization leads to risk minimization. In a cluster it is easier to keep a record of what people are doing, and in China this means keeping a rein on artistic sentiment.

In addition, such clusters are also providing opportunities for emerging business practices, which in the international community might be considered rent-seeking. An example of this is found in another arts village, this time at the other 'high-value' end of the creative continuum. Songzhuang Village in Beijing's Eastern Tongzhou District is the epicentre of China contemporary arts scene. Here, more than 2000 artists produce original work. Many of these paintings make their way into the thousands of contemporary art galleries in China's big cities. In addition, tax is levied on paintings sold in exhibitions and auctions, which adds kudos to the career trajectories of local officials, even if the tax revenue goes directly to the national and the city governments. Yet despite the invocation to be creative, many artists are choosing to opt for what the market wants: in this case it is the tourist market.

The economic logic of clustering, taken from the work of Michael Porter (1998), has been distilled into assorted creative parks, bases, incubators, industrial districts, creative cities, and creative regions. The idea of creative clusters makes sense in today's China, given the legacy of collective organization: the People's Communes (1950s–1960s), the town and village enterprises (TVEs) (1980s–1990s), the science and technology parks (1990s–2000s), and the media conglomerates (*jituan*) (instigated in the late 1990s–early 2000s). In differing ways these collective models responded to social and economic reforms. The common ingredients were a high degree of hierarchical management, favourable investment policies, and state supervision.

As mentioned above, initial enthusiasm for creative industries came from Shanghai. On 8 January 2005, the Shanghai Creative Industry Centre (SCIC) began operations. A month later the Shanghai Economic Commission published a glossy promotion called *Shanghai Creative Industry Clustering Parks*. The 14 clusters in this 'first wave' were mostly disused industrial spaces in high-value commercial districts. Some were already operational. The momentum was soon followed by other cities. In December 2006 the Beijing city government formally established its first wave of 'cultural creative clusters'.

The compromise term 'cultural creative industries' was taken up in Beijing. Beijing's first wave of 10 cultural creative clusters was announced amid great fanfare in December 2006. These included the resolutely uncreative Panjiayuan 'flea market' (a market for antiques and replica antiques), the 798 Art District at Dashanzi, and potential 'big picture' developments such as the Zhongguancun Creative Industries Pioneer Base, the Songzhuang Art and Cartoon Zone, the Huairou Film and TV Base, and the Beijing Cyber-recreation District in Shijingshan.

While Shanghai had listed some 70 creative cluster sites on its books by 2007, Beijing aimed to emphasize key projects. However, by 2008, the number of projects in Beijing claiming, or at least aspiring, to be clusters, zones and cultural creative parks had suddenly risen. These included theme parks, folk-custom streets, a fashion business district (FBD), Olympic Games constructions, and sports culture centres.

Outside Beijing, meanwhile, the creative space momentum was building. To the south-east of Beijing, the city of Tianjin had undergone massive infrastructural development since 2005. In the shadow of Beijing, as far as the benefits of the international tourist industry were concerned, Tianjin's planners believed the city could share Beijing's resources. Tianjin was competitive in the sense of having lower business and living costs. These might be attractive elements for wooing creative enterprises and talents. In the north-east, Dalian was positioning itself in the creative industries, opting to becoming the new Bangalore of East Asia. In the south, Shenzhen was aiming to exploit Hong Kong money, utilizing its lower production costs and its migrant workforce. In the south-western centre of China, Chengdu and Chongqing were busy with their own creative city and creative industry plans, leveraging on the national policy of revitalizing the western regions. For city planners in these huge cities, the creative industries were a new development format in which they could inscribe regional characteristics and provide an outlet for the increasing creative workforce. In the more coastal regions of Hangzhou, Suzhou and Nanjing the focus was animation, software, as well as the reuse of industrial space for housing a new wave of artists, designers, and related professional services. Hangzhou, in particular, has set its aspirations to become the 'Silicon Valley in Paradise', obviously a brand strategy to compete with China's other Silicon Valley wannabes Zhongguancun (Beijing) and the Zhangjiang Hi-tech Zone in Pudong (Shanghai) (see Zhang et al. 2007).

The development of cultural and creative clusters in China has advanced through several distinct, although overlapping, stages. The first stage saw a sudden outbreak of specialist agglomerations, for instance spaces dedicated to industrial design, antiques, jewellery, animation, painting and sculpture. Many of these designated industry clusters were situated in disused industrial space. The idea here was less about linking with the tourism market and more about providing resources and a workspace for people with similar skills.

Meanwhile, a second phase of clustering projects saw the evolution of artist zones and cultural districts, essentially organic developments combining strong tourist pull with consumer services. Certain districts within large cities provided a milieu in which artists and media workers could enjoy a greater sense of freedom to express themselves. Examples included Beijing's Nanluo guxiang hutong precinct, the Songzhuang Art District and Shanghai's Taikang Rd. Situated close to art colleges, galleries and media schools, and with historical legacies, these districts attracted international designers, writers and artists, which also produced an informal economy of coffee, pasta and local beer.

A third expression of the creative cluster can be termed the 'related variety' model. Many artists and media producers were making use of reconverted factory space. The key difference with the specialist cluster mentioned above is a mix of small enterprises specializing in design, media production, fashion, painting, photography and sculpture. The related variety model has become the default setting for local governments keen to exploit the link between art and tourism. In Beijing, the most well known art space is 798, the reconverted switching factory in the area known as Dashanzi (Keane 2009). Not far away is Caochangdi, another emergent art space, which is absorbing spillovers from 798, which many feel is already too commercialized. In the north Chinese city of Dalian in Liaoning province, the Xinghai Creative Island claims to be Dalian's own 798. Inside the reconverted factory situated next to a 'thinker's park' (*sixiang gongyuan*), the walls are emblazoned with the UK definition of creative industries, listing the various sectors. The chosen few – the space is relatively small compared with the expanse of 798 – occupy pristine new workspaces. Offerings include producing oil painting, graphic design, media content, animation, porcelain, and reproductions of traditional artefacts from the Shang Dynasty. In addition, there is even a space specializing in nude photography. Not to be outdone, nearby Qingdao in Shandong province offers the aptly termed Creative 100 precinct. In Shanghai, the 1933 Old Mill Factory is a reconverted abattoir that showcases design, fashion and new media. Other related variety examples in Shanghai are Tianzifang, the No 8 Bridge, the Media Culture Park, and the Modern Industry Mansion Park. Hangzhou is represented in this format by Loft 49 and the A8 Art Commune, while nearby Nanjing has Nanjing 1912 and the Creative East no 8 District. In South China, Shenzhen has the OCT Loft, Shenzhen and the F518 Creative Fashion Park, while Tianjin has established the No. 6 Warehouse, the Hualun Creative Factory and the Lingao Creative Industries Park. In Chongqing, the Tank Loft is a reconverted munitions factory once used by Chiang Kai-shek (Jiang Jieshi). In southern Foshan in Guangdong province, the Cultural Creative Park is a large-scale development that combines elements of various media and folk cultural productions.

In general, however, these creative industries cluster projects have been led by real estate developers. Gentrification, together with consumer service functions, has served the bottom line. In all, hundreds of reconverted factories throughout China have incorporated assorted design, painting, media, fashion and advertising services, which are made more commercially viable by recreational add-ons – bars, restaurants, massage, book and souvenir shops. In a sense, it is not the creativity or the networks of interaction that fund this wave of construction: it is the production and sale of tourist commodities. While one could argue that this mix of design-related activities does constitute 'related variety', the commercial focus has conspired to produce competition for markets rather than cooperation in learning. As a result, many have bemoaned the crass commercialization and loss of authenticity. The end result has been an increase in land value and rents.

A further stage has seen the formation of media content clusters within existing industrial zones throughout China, notably in the field of animation, and mostly concentrating on outsourcing. As mentioned above, the industrial park model is another important dimension of China's current soft power push. Within this vision, animation and related digital content are key targets. Since 2005, China has established 17 accredited animation bases. The question of why there needs to be 17 national animation bases would seem to be a valid one for an outside observer. However, in China, the conferral of national status is much sought after. The main centres, often situated within an existing industrial area, are in Shanghai, Beijing, Hangzhou, Suzhou, Shenzhen, Dalian, Suzhou, Changzhou and Wuxi. Local governments offer a range of industry sweeteners, such as preferential policies enabling start-up firms to enjoy tax holidays, to obtain housing and educational services for employees and their children, as well as financial incentives if content is successful. In the main, this entails content being purchased by CCTV.

The problem with the industrial park model is location. Often located outside the CBD in the fringe industrial zones, the centres are not able to tap into the dynamism of the very consumers they are targeting. As one digital content executive interviewed commented: 'Such parks are for nerds'.[1] Another successful animation company CEO suggested to the author that it was fine to be in an industrial park as far as the business incentives were concerned but it was not a place to be creative. In short, much of the activity in the industrial parks is routine fee-for-service work. In this model, moreover, the transfer of knowledge from international companies is limited. Foreign studios outsourcing in such locations continue to resist handing over control of creative work (e.g. conceptualization and pre-production).

On a larger industrial scale, stand-alone cinema and television production centres began to appear, servicing the domestic audio-visual market, engaging in co-productions, or outsourcing production from Taiwan, Korea and the United States. To compensate for the cyclical nature of audio-visual production, these centres offer a theme park function which cashes in on the success of cinematic and TV drama output. In the north of Beijing, the Huairou Film production centre has established itself in recent years in competition with more established film studios. The largest film park, however, is Hengdian World Studios in Zhejiang province, where tourists can see a re-enactment from the movie *The Opium War*, and be offered a tour of the set of *Hero* (Zhang Yimou) and *The Emperor and the Assassin* (Chen Kaige). The landlocked nature of Hengdian puts it at a disadvantage to the large urban centres of Shanghai and Beijing. To offset this, Hengdian absorbs a great deal of low-cost television drama production, particularly dynastical costume dramas.

Finally, yet another model is the incubator model, often with a purported emphasis on R&D, and often with the declared intention of making science parks more 'creative'.

The proximity of many science and technology (S&T) parks, also called innovation parks, to prestigious universities and development zones reflects a national desire to incubate something above and beyond standard products. The problem with this vision is that many existing S&T parks are hardly innovative. Critics like Wang Jici (2007) contend that industrial parks are regarded by government as infrastructural, a means to attract enterprises from outside. In contrast, the industrial cluster, or the creative cluster model, is about enabling a learning economy within. Proclaimed 'creative incubators' are now to be found in Chongqing (the Ideas Industry Centre); Tianjin (the Heping District Creative Animation Park; the Taida Science and Technology Park); Dalian (the Creative Incubator Garden); Hangzhou (the Hangzhou Innovation and Creative Industry New Base); Beijing's Zhongguancun Creative Industries Pioneer Base; and Shanghai's Zhangjiang Hi-tech Zone in Pudong and the KIC (Knowledge Innovation Community) in Yangpu District (Zhang et al. 2007).

Concluding remarks: What does this mean?

China's great experiment with cultural creative industries is something that has so far escaped mainstream academic attention. When the Chinese media are celebrated in the cultural studies and post-colonial studies traditions, it is typically the works of filmmakers who offer a version of Chinese culture that conforms to Western stereotypes or dissident writers who perpetuate a pessimistic image of Chinese society and the state's unremitting political control. Political economy of the media offers a more negative version of China's positioning, subscribing to a cultural domination model in which the control of world communications emanates from a centre – the US capitalist dream factory.

While the political economy of the media does acknowledge that China is changing, many working in this critical tradition worry that it is becoming too globalized, in turn reducing diversity and ushering in too much 'inappropriate' global culture – Starbucks, Hollywood blockbusters and Hullo Kitty. As this paper has shown, cultural adaptation has occurred throughout the past, and continues to do so. China takes in international models and development formats. As culture is attached to economic development we can see the totalizing model of the command and control centre shifting to regional competition and specializations. The key example is the creative cluster. Whether this massive development of regional clusters, bases, zones and precincts will prove successful in brokering 'national' independent innovation, over and above luring tourists, remains to be seen.

What is evident, however, is that new forms of social relations have emerged between propaganda officials and developers, between propaganda officials and artists, media producers and entrepreneurs. For local officials, too, there is much at stake: if the development project brings economic benefits they stand a chance of monetary gain and career advancement. Developers have moved into the picture, seeing creative clusters as the opportunity to offer new master plans, often complete with shopping malls. For this reason, the creative industries, with their unapologetic link to wealth creation, are a development model to be taken seriously. For artists and other creatives, too, there are winners and losers. There is a great deal of interest in how to brand and market oneself, how to respond to the latest fashion, how to develop one's media profile, and how to target the tourist market. In this, some degree of authenticity is disappearing.

Note

1.   Jerry Wang, CEO Goyoo Media, interview with author 22 July 2006.

References

Caves, R. 2000. *Creative industries: Contracts between art and commerce.* Cambridge, MA: Harvard University Press.

Chang, S. 2008. Cultural creative industries. *Urban China*, no. 33, December.

DCMS. 2001. *Creative industries mapping document.* London: DCMS.

Hu, Jintao. 2007. *Hold high the great banner of socialism with Chinese characteristics and strive for new victories in building a moderately prosperous society in all respects.* Report to the Seventeenth National Congress of the Communist Party of China, Hu Jintao, 15 October. English version available at http://www.bjreview.co.cn/document/txt/2007-11/20/content_86325.htm (accessed 8 February 2008).

Keane, M. 2002. Facing off on the final frontier: The WTO accession and the re-branding of China's national champions. *Media International Australia* 105: 130–47.

———. 2007. *Created in China: The great new leap forward.* London: Routledge.

———. 2009. The capital complex. In *Creative economies, creative cities: Asian–European perspectives*, ed. Lily Kong and Justin O'Connor. New York: Springer.

Kong, L., C. Gibson, L.-M. Khoo, and A.-L. Semple. 2006. Knowledges of the creative economy: Towards a relational geography of diffusion and adaptation in Asia. *Asia Pacific Viewpoint* 47, no. 2: 173–94.

Liu, Lydia H. 2004. *The clash of empires: The invention of China in modern world making.* Cambridge, MA: Harvard University Press.

Porter, M. 1998. Clusters and the new economics of competition. *Harvard Business Review* 76, no. 6: 77–90.

Rosen, Stanley. 2002. The wolf at the door: Hollywood and the film market in China. In *Southern California and the world*, ed. E.J. Heikkila and R.E. Pizarro-O'Byrne, 49–78. Westport, CT: Praeger.

Ryan, Bill. 1992. *Making capital from culture: The corporate form of capitalist cultural production.* Berlin: Walter de Gruyter.

Sigurðsson, Geir. 2008. Is there such a thing as Asian creativity? Paper presented at the Asian Creativity in Technology and Culture conference, 12–16 November, Trondheim University, in Trondheim, Norway.

Spence, J.D. 1974. *Emperor of China: Self-portrait of K'ang-hsi.* New York: Knopf.

Wang, J. 2007. Industrial clusters in China: The low road versus the high road in cluster development. In *Development on the ground: Clusters, networks and regions in emerging economies*, ed. A. Scott and G. Garofoli, 145–64. London: Routledge.

Zhang, Jincheng. et al. 2007. *The Blue Book of China's creative industries.* Beijing: Beijing Economic Publishing.

Zhao, Yuezhi. 2002. Enter the world: Neo-liberal globalisation, the dream for a strong nation, and Chinese press discourses on the WTO. In *Chinese media, global contexts*, ed. Chin-Chuan Lee, 32–56. London: Routledge.

Adapting the mobile phone: The iPhone and its consumption

Gerard Goggin

Adapting the mobile phone: The iPhone and its consumption

Although a mere three decades old, the cellular mobile phone is an important location of contemporary culture and its reproduction and variation. In this paper, I wish to consider cultural adaptation through a discussion of the mobile phone. My case study is the Apple iPhone. Introduced in mid-2007 to much acclaim, the iPhone is an excellent example of adaptation because it is explicitly conceived as an intervention into the styles and genre of contemporary culture – notably mobile phone cultures, Internet cultures, and the broader scenes of digital culture, and what it represents for cultural transformation in general. Moreover, the iPhone and the terms of its introduction have put strong emphasis on the active roles that people play in orchestrating and using mobile, digital cultures.

The cellular mobile phone was first and foremost an adaptation of the telephone. This process itself spanned the best part of the twentieth century, drawing together various complex revisions: the reworking of radio technologies and radio spectrum; the remediation of the telegraph; the reimagining of mobility; and the acoustical recrafting of voice telephony for the portable instrument. The 1980s is broadly the period in which the classic form of the mobile phone was stabilized. By this I mean that the decisive shift to a stand-alone portable telephone in this decade of the first-generation analogue mobile phone provided the material basis for a set of new affordances and design features that are now regarded as standard for a cell phone (Goggin 2006). In the 1990s, when the second-generation digital mobile system took over, the cell phone became smaller, more portable, more domestic and closer to the body (Fortunati 2006), and this was accompanied by the inclusion of new features, capabilities and communicative architectures, as well as cultural expectations and routines, into this pocket-sized technology. In the 1990s, the cell phone became part of everyday life, and loomed large in the conjuring of those involved in design and fashion, and entertainment and media, as much as the worlds of telecommunication.

If the telephone was a rich repository of adaptation, the mobile phone greatly intensified and accelerated this. Further, with the advent of the cell phone, the importance of telephonic metamorphosis to cultural adaptation in general became vastly more significant. Although the mobile phone is an important site of contemporary culture – as explored in the special 'Mobile Phone Cultures' issue of *Continuum* 21, no. 2 (2007) – it has not yet been viewed systematically through the framework of adaptation central to this issue and developed elsewhere (e.g. Hutcheon 2006). Michael Keane and Albert Moran's work has richly theorized the processes of adaptation in television and film formats, genre, institutions and structures (Keane, Fung, and Moran 2007; Moran 1998; Moran and Keane 2007). For instance, in their examination of fashioning television program formats in Asia and the Pacific, they propose the present as an era of abundance, and ask:

> What then is the motor or source of this differentiated abundance? How does it register as a phenomenon and how does it come about? The significant dynamic of the present era in television seems to be one of adaptation, transfer and recycling of narrative and other kinds of content. The phenomenon is widespread even while the particular term or set of terms that cover its operation are quite imprecise and slippery to apply. Although we invoke a series of labels including adaptation, transfer, recycling, translation, remaking, spin-off, and re-versioning, we recognise that this kind of activity needs greater commonly agreed upon terms. (Keane and Moran 2005)

As a way to open up the questions of mobile phones and adaptation, it is useful to think about the extent to which mobile phone cultures are involved in broadly similar dynamics to television or other media forms – and what the major differences might be. Mobile content, including narratives, programs and content, is now becoming important to adaptation in this sphere, but this is only part – and so far a minor part – of what is at play here and in the larger dynamics of culture in which the mobile is set.

To start with, the mobile is an everyday device, object or thing (Turkle 2007). The materiality of the mobile phone is very important to how it is understood and how it bears, marks and is grasped by culture. Further, the mobile has been a much-copied and adapted object. There is no comprehensive history or account of these adaptations of which I am aware. In the absence of this, I would point to the new kinds of adaptations around the world, but especially those occurring in the Asia-Pacific. These include the fertile and resilient career of text messaging; the cumulative imaginings of micro-adaptations of users and user-generated content, such as those charted by Larissa Hjorth (2009); the quite different mobile media adaptations emerging in Korea and Japan; and the large-scale industrial cloning, copying and reinvention of mobiles by new social groupings, and national formations (such as those represented in China's stark emergence as a massive mobile-using and manufacturing country).

Two important terms used to discuss mobiles have been 'design' (regarding which there is a large and diverse literature) and 'customization' – and bringing these to bear on the cultural adaptation framework would be a fruitful process. To bring the mobile phone into discussions of adaptation also highlights the importance of technology. Notoriously difficult to define, the aspect of technology is crucial to grasp for understanding mobiles – as it is for grappling with the new character of globalization and culture. Here I am interested in technology and adaptation, which is raised – though not necessarily explicitly – in the literature on global technologies (Law, Fortunati, and Yang 2006).

In this article I can only aim to make a preliminary contribution to this larger project of thinking about mobiles and adaptation. Rather than focusing on the myriad adaptations of the 1970s, 1980s and 1990s, I wish to concentrate on a more recent and highly publicized effort at remaking the mobile: the project of the iPhone. For a number of reasons, the iPhone is a very interesting case of mobile phone adaptation. First, the iPhone does not present itself as an adaptation: Apple professes to be utterly changing the mobile – at least in its promotional literature, but also in other important ways. Second, the kind of adaptation the iPhone represents is about adapting the mobile phone for the Internet. It is about adapting the mobile to finally put it at the centre of computing, the Internet and digital culture. For instance, the iPhone is very much a platform for, and creature of, its applications. Its 'Apps Store' allows a wide range of programs and applications for the iPhone to be readily purchased and downloaded, bringing the affordances of computing to the mobile phone in a form that previously has not been possible. Third, the iPhone is very much a haptic adaptation. The mobile phone very much emerged as a haptic technology, rather than primarily an aural, listening or speaking technology, especially with the popularity of text messaging – underscored by the emphasis on the mobile as a 'hand' (or 'handy') technology, and especially through the text messaging 'thumb' culture. Fourth, the iPhone promises to make the mobile even more customizable and adaptable – identity on the move, made to order. For instance, the iPhone menus can be configured to a much greater extent than previously in line with the user's wishes: thus each iPhone screen can look quite different.

In what follows, I want to examine the sense in which the iPhone works as a cultural adaptation. My argument is threefold. First, I suggest that the iPhone is indeed an important adaptation of the mobile phone – though not quite in the terms in which Apple presents it. The iPhone does draw from the grammar of mobile phone design, and combine and rework a number of well-established affordances, elements and technologies. And it also borrows from the well-established, distinctive traits of Apple 'i' technologies: the hardware of the iPod, and the software, and intellectual property and digital rights management regime of the iTunes application. Thus the iPhone pushes the mobile much more towards the world of computers and the Internet. Second, in its consumption, the iPhone sees the mobile phone take on even more and quite novel everyday uses and meanings. Third, because of its new affordances and design, and also the larger scene of its everyday invention and consumption, the iPhone sets in train a new logic of adaptation that Apple cannot license. With the iPhone, Apple sets out take the control of adaptation from the hands of mobile phone carriers and manufacturers, and to allow new flexibility, third-party applications, programming and data exchange. Apple tries to control, circumscribe and manage the way in which it puts new powers of adaptation in the user's hands: to allow the user to get under the bonnet, to borrow a motoring metaphor. What Apple does not reckon on, or at least cannot hold at bay, is the warp-speed way that unauthorized modification of the iPhone occurs – thanks to enthusiastic hackers and the immediate, sweeping dissemination of hacking tools via the Internet. Certainly, the bulk of users still may find such do-it-yourself adaptation of the iPhone too difficult, but a larger-than-usual group of users did avail themselves of the iPhone hacks. All of which probably just adds further to the aura of the iPhone.

The beginnings of the iPhone

The first time a technology came to be called the iPhone was actually in the mid-1990s, when it meant the 'Internet phone'. With the rapidly growing mass consumption of the Internet, developers were hard at work to devise a form of telephony that could work via the new network of networks. This form of Internet telephony was called the iPhone. This has now developed into a relatively easy-to-use household technology called Voice over Internet Protocol (VoIP), with its best-known proponent being Skype. In the 1999–2000 period, a different class of mobile phone devices was marketed, also bearing the name of the iPhone. Key to these was a claim that the mobile phone would now become a prime device for accessing the Internet.

For its part, Apple started work on its iPhone in 2005, with a prototype finally emerging in mid-2006 (Vogelstein 2008). Previously, Apple had already created one mobile phone, aimed at preserving its hold on the digital music market. This was the ROKR, released in September 2005. The ROKR was a joint venture between Motorola and Apple. Motorola had responsibility for the phone (in conjunction with the mobile carrier Cingular), while Apple focused on the music software (Vogelstein 2008). Unlike Motorola's sleek, popular RAZR phone, the first version of the ROKR looked a lot like a classic, rather clunky cell phone. It only held 100 songs, and did not allow music to be directly downloaded. However, the ROKR was the first mobile phone to feature the Apple iTunes application (see Figure 1).

The latest version, the ROKR E8, launched in April 2008, does much more resemble the RAZR (see Figure 2). Nonetheless, its reputation was dismal, the apogee of its reception being *Wired* magazine's cover, with its headline: 'You call *this* the phone of the future?' (Vogelstein 2008).

At the time the ROKR was invented, the talk was of the 'music phone'. However, a key problem with the collaboration between Apple and Motorola in devising the ROKR was over the computer giant's approach to intellectual property and digital rights management:

> A key part of the iTunes package, for example, is FairPlay, Apple's digital rights management software ... FairPlay would set limits on the new phone: It couldn't play music from any major online store but iTunes. It couldn't hold more than 100 songs. 'It's obvious why Apple is doing this,' says Patrick Parodi, head of the Mobile Entertainment Forum, an industry trade group. 'They don't want to cannibalize the iPod.' (Rose 2005)

As the ROKR went to market, the iPhone was in development – allowing Apple tighter control over music and other content. Clearly, Apple's strong suit lay in its strengths in computers, operating systems and integrated suites of applications. So, for instance, its engineers rewrote the Apple OSX operating system for the iPhone. Apple also shone in the area of design, with its iPhone building on the classic shape and look of its classic iPod device. However, Apple showed less proficiency with the other features of mobile phones.

Figure 1.   The ROKR.

Figure 2.   The ROKR E8.

The much-heralded iPhone ran initially on the 2.5-generation digital mobile phone network, with the third-generation (3G) iPhone following roughly a year later, in mid-2008.

'Biggest launch since the Apollo program': Making the iPhone fever

> It was the biggest launch since the Apollo program. How did Apple's smartphone – which slickly packages features already available in other handsets – become such a highly anticipated phenomenon? The answer lies not in Steve Jobs' (undisputed) marketing prowess but in the abject failure of other handset manufacturers to deliver a portable Internet device with mass appeal. So the iPhone has ascended, and its liftoff was a rousing success. (Geekipedia 2007)

Perhaps what most distinguished the iPhone from the many other adaptations of mobile phones was its rapturous reception and, hand in hand with this, Apple's phenomenally successful marketing campaign. Herein lies the paradox of adaptation that the iPhone represents. The iPhone is clearly an adaptation of the mobile phone. As *Wired* magazine's Geekipedia points out, the iPhone is an obvious descendant of the smart phone – the multimedia mobile phone that combines various computer programs with entertainment options. Yet the 'biggest launch since the Apollo' rebadges this evolution as a revolution. Clearly, design values, chic and usability – all features of the Apple world of iMac and iPods – are key to this. So too is the careful crafting of the iPhone's reception by its makers.

The arrival of the iPhone was much anticipated, and featured long queues of aficionados, who were prepared to wait out in all conditions for the privilege of being the first to hold and try the new device. The iPhone queues started with the first launch in the

Surrounded by cheering Apple Store employees, one of the first iPhone buyers leaves the
store on Fifth Avenue in New York,.
Photo: *Reuters*

Figure 3.   'Frenzy as iPhone go [sic] on sale'. Source: *Sydney Morning Herald*, 30 June 2007
(www.smh.com.au).

United States in late June 2007 (see Figure 3). Tech blogger Caroline McCarthy provides
typical coverage: 'Check out my grainy little video – that's the iPhone line in midtown
Manhattan. Can you believe that there's apparently barely anyone waiting at the city's
AT&T stores?' McCarthy actually plotted the length of the queue in midtown Manhattan
at 3.00 p.m. on 29 June 2007 on Google maps (McCarthy 2007).

After its US debut, Apple staged the launch, complete with queues, in many
subsequent countries – each launch not only a local media event, but often attracting
international attention also.

In July 2008, I happened to be in Wellington, New Zealand when one morning the
arrival of the iPhone, complete with now customary queues, was announced. Not only did
this signal the arrival of the iPhone in Aotearoa but it also marked the launch of the first
3G iPhone worldwide. The news of the launch in Wellington was quickly updated with the
revelation that the young student at the head of the queue, who was the first person to
obtain an iPhone, was actually paid to do so. It has now been acknowledged officially
elsewhere by at least one operator that the iPhone queues are a key marketing ploy, and
that extras are paid to participate:

> Poland's biggest telecoms operator, Telekomunikacja Polska, acknowledged last week that it
> had paid young, hip-looking film extras to stand in queues for the national launch of Apple's
> iPhone. 'It was a marketing move. We thought it was a pretty interesting strategy,'
> TP spokesman Wojciech Jabczynski said. 'The aim was to attract attention. The people in the
> queues told passers-by about the iPhone.' (*The Australian* 2008)

The marketing hype not withstanding, the iPhone was named the invention of the year by
Time magazine in 2007:

> The thing is hard to type on. It's too slow. It's too big. It doesn't have instant messaging.
> It's too expensive. (Or, no, wait, it's too cheap!) It doesn't support my work e-mail. It's locked
> to AT&T. Steve Jobs secretly hates puppies. And – all together now – we're sick of hearing
> about it! Yes, there's been a lot of hype written about the iPhone, and a lot of guff too. So much
> so that it seems weird to add more, after Danny Fanboy and Bobby McBlogger have had their

day. But when that day is over, Apple's iPhone is still the best thing invented this year. (Grossman 2007)

Grossman gives five reasons for his verdict: the iPhone is 'pretty', it is 'touchy-feely', 'it will make other phones better', 'it's not a phone, it's a platform', and it 'is but the ghost of iPhones yet to come' (Grossman 2007). The accompanying photo of the device is captioned: 'It's a genuine handheld computer, the first device that really deserves the name.'

iPhone apps: New forms of mobile consumption

Mobile phones, even the application-rich smartphone, have been difficult devices for many to manipulate, and to reprogram in the way that users of computers, and especially users of the Internet, expect to be able to do. There is an irony in this. As mentioned, mobiles have actually been an eminently customizable device in another way. Users have adorned them with their favourite keepsakes. They have changed the faces and colours of their cell phone. They regularly change the ringtone, screensaver or desktop. And they care intensely about the mobile as a signifier of fashion and identity.

However, thinking about and using the mobile phone as something that can be programmed and networked, according to the user's preferences, has been a difficult proposition. Moreover, there has been little of an open market in mobile phone applications at the consumer level. The typical scenario is that applications – notably games – can be downloaded via mobile Internet (WAP) sites, or via premium mobile content, and then can run on the device (memory permitting). There are also many Internet websites that offer applications for mobiles. Certainly there are some mobile users who do regularly download such applications (evidence of a burgeoning mobile content industry) – but the process is not especially user friendly.

Enter the iPhone 'Apps store'. Using the iTune interface and user experience, the Apps store does make it much easier to be aware of, choose, pay for and download applications for iPhone. Apple's pitch is: 'Applications unlike anything you've seen on a phone before':

> Applications designed for iPhone are nothing short of amazing. That's because they leverage the groundbreaking technology in iPhone – like the Multi-Touch interface, the accelerometer, GPS, real-time 3D graphics, and 3D positional audio. Just tap into the App Store and choose from thousands of applications ready to download now. (Apple Apps store, www.apple.com, 15 October 2008)

Both via the Internet and using the iPhone itself, the experience of finding applications is much enhanced. Not only is the iPhone a signal adaptation of the Internet and mobiles, it is highly adaptable by its users. The applications and programming options of the iPhone themselves feature very visibly in iPhone culture, as the Apple promotion suggests – as users try, swap and discuss applications. It has also meant that the iPhone is an important new platform for developers, a community which has often found the experience of developing applications for mobiles a frustrating experience.

Indeed, the iPhone has faced serious criticisms from developers. In the first place, Apple launched the iPhone without allowing access to third-party developers. This allowed it to announce the release of a software development kit with some fanfare. The basic terms upon which Apple engages with iPhone application developers are still quite controversial, and are seen by many as too restrictive and slanted in Apple's favour. There have been a number of celebrated examples in which Apple has cracked down on developers.

With Apple's easing of the restrictions on developers, the applications available have often centred on novel uses of a mobile phone. A number of these revolve around the iPhone's three-element accelerometer. The iPhone's accelerometer is a sensing device that is able to gauge the orientation of the phone, and make appropriate changes in the screen. For instance, someone viewing photos on their iPhone can rotate the device 90 degrees, from portrait to landscape layout, and the display will detect the movement and change accordingly (Apple 2008). The iPhone is equipped with two other sensors: a proximity sensor, and an ambient light sensor.

Rich Ling opens his book *New Tech, New Ties* with a story of a plumber who knocks on his door, distractedly talking on the phone (Ling 2008). When Ling opens the door, the plumber, without so much as a by your leave, walks in, continues his conversation, and begins measuring up for the job. With the iPhone, the mobile phone is not only the great communication and business tool of the tradesperson, it is literally a tool they can use in their work. With the application 'Level', the iPhone becomes a spirit level (or carpenter's level), which can be used to see whether a surface is flat or plumb. The spirit level was apparently invented by Melchisedech Thevenot around 1661 (Wikipedia 2008), so the iPhone amounts to a nice reprise almost four and a half centuries on. There are now a myriad of uses of the iPhone's new adaptation of sensing technology, including applications that allow you to play games swinging the phone, such as iBowl ('Simply swing your iPhone like a bowling ball and see how many strikes you can get'). Here, the iPhone is clearly an adaptation of the gaming practices and moves familiar from Nintendo's Wii remote, the wireless controller for the popular video game console (Johnson 2008).

There is much more to say about the burgeoning culture of the iPhone centring upon its great potential for adaptability with downloading of apps, flexible configuration and new logic of sensing, motion and touch. The iPhone is a novel *combinatoire*, at the least, for melding mobile, computing and Internet cultures. But what are the limits and politics of adaptation that its architecture, code and design allow and constrain?

'It will make other phones better': iPhone's innovation

If we discount the most overblown claims for the iPhone – the 'Jesus' phone, the most awaited launch since Apollo 11, and so on – then the central claim concerning the iPhone is that it strikes a fatal blow to the locked-down mobile phone platforms. In the words of *Wired*'s Fred Vogelstein, the iPhone 'blew up the wireless industry'. Whether in these terms or others, this is the central claim about the magnitude and meaning of the iPhone's killer adaptation. While conceding various shortcomings of the iPhone, Vogelstein submits that:

> The iPhone cracked open the carrier-centric structure of the wireless industry and unlocked a host of benefits for consumers, developers, manufacturers – and potentially the carriers themselves. Consumers get an easy-to-use handheld computer. And, as with the advent of the PC, the iPhone has sparked a wave of development that will make it [the mobile phone] even more powerful ... Manufacturers will have more control over what they produce; users – not the usual cabal of complacent juggernauts – will have more influence over what gets built. (Vogelstein 2008)

This will result in the scales being lifted from the eyes of Apple, for the common good:

> wireless carriers begin to show signs of abandoning their walled-garden approach to snaring consumers ... Eventually this will result in a completely new wireless experience, in which applications work on any device and over any network ... (Vogelstein 2008)

On this view, the benefits of the iPhone stand not only to accrue to Apple and to iPhone application developers but also to the mobiles industry itself (what in the United States is typically referred to as 'wireless'):

> It may appear that the carriers' nightmares have been realized, that the iPhone has given all the power to consumers, developers, and manufacturers, while turning wireless networks into

dumb pipes. But by fostering more innovation, carriers' networks could get *more* valuable, not less. Consumers will spend more time on devices, and thus on networks, racking up bigger bills and generating more revenue for everyone. (Vogelstein 2008)

The iPhone, then, is an adaptation of the mobile phone highly suited to the master themes of its time: end-to-end digital innovation.

There is some truth to this account. Apple was able to negotiate unusually favourable terms with carriers, regarding the split of revenues. This was one reason why Apple also struck exclusive deals with carriers – such as its deal with AT&T when it first launched in the United States. While few details are in the public domain, it does appear that Apple did manage to gain a better deal than that typically struck with other mobile content providers (the bargaining power of the carriers being a great bugbear in the industry). However, it is important not to overstate this achievement. The political economy of mobile media is still very much structured and controlled by the cellular mobile carriers, which by virtue of their control of the networks, custody of the customer databases and long-established sunk capital and pervasive presence still command what is otherwise a maelstrom of media convergence.

It is important to view the emergence of the iPhone, and Apple as a significant player in the cultural and political economies of mobile media, from the consumer's perspective. Apple is a peculiar kind of cell phone manufacturer, enjoying compelling and desirable horizontal integration. Thus it may be able to extract better gains from the negotiation with mobile carriers. Even so, there is no reason to think that Apple will pass on these gains holus bolus to the consumer or end user. To the contrary, there is a lively discussion and considerable literature on the strangleholds that Apple places on consumers of its computers, software and digital content. These strictures might be different from the suffocating embrace of the mobile carriers, and the norms of use and standard contracts that obtain there – and it is heartening to see the mobile carriers' systems of control challenged by the iPhone. However, Apple has its own well-defined interests in not opening things up too much; indeed, the company consistently seeks to circumscribe use and circulation of digital materials, and to demarcate the arena of digital culture in new ways, that serves its own interests. It is this aspect of the iPhone's adaptation of mobiles that goes squarely to the raging battles in digital cultures that I will now consider in some depth.

The cultural politics of iPhone modification

The flipside of the iPhone fever, and its Jesus-phone status, has been the clamorous role that hackers, with their devilish intentions, have played in its reception. Here, unseen and undesired adaptations have eventuated, despite the company's best wishes and attempts to control unauthorized uses. Recall that, when first launched, the iPhone was very much a 'closed' device at a number of levels: locked to one network provider; its applications only provided by Apple, and not open to the wider developer community; and its content safeguarded by the digital rights management regime well established through iTunes. Yet no sooner had the iPhone launched than it was hacked, with modifications and instructions freely available on blogs and websites (for instance, see Apple Insider 2008 for documentation on the hacking of the New Zealand 3G iPhone immediately after its launch).

While there was talk, and some evidence, that Apple did take legal or corporate action to stop or at least deter users from unlocking or modifying the iPhone, it could be argued that these unauthorized adaptations, displaying such keen interest in laying it bare, played into the mythos of the device. While many devices and software are now routinely cracked and modified, there was a visibility to the hacking of the iPhone that made the usually more underground tools and downloads much easier to find. Many users in countries such as Australia, in which the iPhone had not been launched, were able to buy the device, activate and connect it with a provider of their choice – so adding to the advance praise of the technology.

In early 2009, the iPhone is now a more open platform, with a set of controls but also better access for developers, and also having fermented a thriving user culture which itself has some ability to modify its software and hardware. However, the iPhone does still remain quite a 'closed' platform. Jonathan Zittrain's early critique remains the most systematic one, even if it is now a little outdated (Zittrain 2008). Zittrain argues that the iPhone and similar moves amount to 'tethered appliances'. Zittrain opens with an image of Steve Jobs' January 2007 Macworld launch of the iPhone. Like Jobs, Zittrain draws a comparison between the launch of the iPhone and that of the Apple II computer 30 years previously:

> Though these two inventions – iPhone and Apple II – were launched by the same man, the revolutions that they inaugurated are radically different. For the technology that each inaugurated is radically different. The Apple II was quintessentially generative technology. It was a platform. It invited people to tinker with it … The iPhone is the opposite. It is sterile. Rather than a platform that invites innovation, the iPhone comes pre-programmed. You are not allowed to add programs to the all-in-one device that Steve Jobs sells you. Its functionality is locked in, though Apple can change it through remote updates. Indeed, to those who managed to tinker with the code to enable the iPhone to support more or different applications, Apple threatened (and then delivered on the threat) to transform the iPhone into an iBrick … Whereas the world would innovate for the Apple II, only Apple would innovate for the iPhone. (Zittrain 2008, 2)

For Zittrain, the iPod is an example par excellence of the 'tethered appliance' that now threatens to change the shape of the Internet:

> People now have the opportunity to respond to these problems [of PC and Internet failures brought on by bad code] by moving away from the PC and toward more centrally controlled – 'tethered' – information appliances like mobile phones, video game consoles, TiVos, iPods, iPhones, and BlackBerries. (Zittrain 2008, 101)

According to Zittrain, 'tethered devices' fail to be generative platforms, and what's more, they are configured to be actively inimical to user experimentation and co-creation:

> These tethered appliances receive remote updates from the manufacturer, but they generally are not configured to allow anyone else to tinker with them … Indeed, recall that some recent devices, like the iPhone, are updated in ways that actively seek out and erase any user modifications. (Zittrain 2008, 106)

Some of Zittrain's criticisms of the iPhone have less force, given the Apple has now released a software developer kit – and that the device can be activated on a wide range of cell phone networks. Yet Zittrain's critique still carries force, given the dialectics of 'open' and 'closed' in the iPhone as device and platform.

The interesting thing about Zittrain's account is that, having invoked the iPhone as problematic, he does not follow through to analyse the politics of the mobile phone. Rather, his focus is on the Internet and computing devices as generative, with Wikipedia and classic Internet decentralized participatory forms, such as the Request For Comment (RFC), as examples of good practice. In Zittrain's thinking, the mobile phone is used as a contrast to such generative Internet possibilities. The mobile is the foil, the 'bad object', that shows the dystopian future of digital culture. In his discussion of Nicholas Negroponte's controversial 'one laptop per child' project – with its giveaway of one hundred million laptops (the so-called XOs) to each child in the world – Zittrain likes its spirit of generativity, while criticizing its lack of attention to education, and also its tethered aspects. With these caveats, Zittrain much prefers the XO project rather than its cell phone alternative:

> The easier and more risk-averse path is to distribute mobile phones and other basic Net appliances to the developing world just as those devices are becoming more central in the developed one, bridging the digital divide in one sense – providing useful technology – while leaving out the generative elements most important to the digital space's success: the integration of people as participants in it rather than only consumers of it. (Zittrain 2008, 240)

What Zittrain overlooks here is that the mobile phone is already widely used in the developed world – in generative and innovative ways, despite its limitations. In fact, the story of the mobile's low-tech user adaptations in developing countries is now becoming an important part of the media cultural dynamics of the rest of the world (see, for instance, Donner 2009).

So Zittrain's position on the mobile, which is shared by other theorists (Benkler 2006), draws our attention to the control architectures inscribed in coveted devices such as the iPhone. However, rather than following through the intricate struggles over mobile media, such a position vests all too much hope in the participatory culture centring on the Internet – and so both over- and under-reads the politics of adaptation occurring with the iPhone. If the iPhone is a highly significant, if controversial, adaptation of the mobile which at the same time draws attention to, and intervenes in, the cultural politics of adaptation, it is not the only makeover in town. There is much, much more going on in the world of mobiles and mobile Internet.

Take, for instance, another much publicized adaptation of the mobile phone: Google's Android. In September 2008, T-Mobile launched its G1, the first Android phone. The gambit of the Android is its reliance on an open-source model. Like the iPhone, the Android seeks to inaugurate a new kind of mobile phone, with a vision, set of affordances, and assumptions that principally come from computing and Internet cultures. Rather than 'just being a phone', it also aspires to be an alternative platform (see Figure 4). In the case of Android, it is a project of the Open Handset Alliance, a consortium of several companies, Google, Intel and Qualcomm, along with Motorola, T-Mobile, Sprint Nextel, and China Mobile (Helft and Markoff 2007; http://code.google.com/android). Android is based on an open-source licence, and is designed explicitly to compete with the other mobile platforms developed by Microsoft, Apple, Nokia, Palm, Research in Motion, and Symbian operating systems for mobiles.

Google's meteoric rise has been widely dissected, especially for its dominance of the contemporary Internet search and its moves to strategically position itself as central to the future of culture – with its digitization project with libraries. Google now joins Apple as yet one more computer and Internet behemoth seeking to shape the future of mobile media. The politics of the deployment of open source by corporations such as Google, Nokia, and others will be interesting to observe – to see where they diverge over time from

One thing I like about Android...

Watch video
Browser Tricks

Watch video
Copy & Paste

Watch video
Reader & Photos

Watch video
Media & Browser

Figure 4. 'One thing I like about Android'. Source: http://code.google.com/android

the kinds of closed networks that corporations typically favour (for an instructive discussion, see Sawhney 2009, 113–14). Not that the large corporations can rest easily as yet: there are many other open-source mobile phone projects underway, such as the Neo FreeRunner Linux-based smart phone running the OpenMoko platform (Beschizza 2007). The Neo FreeRunner phone went on sale in mid-2008, but is still restricted to developers rather than general consumers (OpenMoko 2008).

Harmeet Sawhney has discussed such developments as mobile extensions of the Internet, looking at their potential for creating what he calls 'arenas of innovation'. According to Sawhney:

> For open-cell-phone enthusiasts, present-day cell phones are like main-frames. The users have little flexibility. They have to function pretty much within the tight parameters set by the system design. The open-cell-phone enthusiasts hope to do to this paradigm what the Jobs and Woz generation did to the mainframes. In effect, they want to make the cell phones as flexible as the personal computers. (Sawhney 2009, 113)

This would presumably be the vision of mobiles, after Zittrain. And it goes to the heart of the fantasy of the iPhone too. But as well as migration of innovation from the Internet (or computing) to the mobile, Sawhney suggests we should pay attention to the possibility of migration the other way (Sawhney 2009, 114–15).

Conclusion

In early 2009, the reception accorded the iPhone has led to various efforts to copy, clone and cope with its success (Fung 2004). Manufacturers from Samsung through LG ('Touch Phones') and Motorola ('Krave' touch-screen phone) to the newer competitor HTC ('Touch Diamond' and 'Touch Pro') took a haptic turn in 2008. And its Blackberry Bold and Blackberry Storm Research in Motion perhaps most explicitly went head-to-head (or digit-to-digit) with Apple (see Figures 5 and 6).

So how might we place these teeming, legion adaptations of the mobile phone? The iPhone is but one project among many that seek to modify the mobile to better take account of the things users expect from Internet and computing cultures, not to mention the genres, forms and practices of convergent media. The iPhone is an especially interesting case but it is such a strong adaptation – indeed, Apple has actively tried to present its device as marking a break with the mobile phone. As I have argued, the millenarian discourse on the iPhone, successful as it has been, obscures the prosaic series of adaptations it involves – as well as a critical evaluation of what is genuinely novel about it.

Figure 5.   'Blackberry Bold vs iPhone 3G'. Source: Boy Genius (2008).

Figure 6.   Blackberry Storm. Source: www.blackberry.com

What is specific about the iPhone is its refiguring of the history, design and habitus of mobile phone culture, and the way in which it moves the mobile much more into the realm of other online media. Yet just as important is the shaping of the iPhone as a device to navigate, arrange and orchestrate everyday life. Here the iPhone is being co-created by a range of other cultural intermediaries than Apple itself, including software developers and users. The third part of the iPhone's new logic of adaptation is the cultural politics of modification, where hackers and battles over code, commons and architecture are playing out.

More research is needed to chart the detailed adaptation of the iPhone across the various places where it has been launched. This was not possible with this paper, but it is an important part of understanding how the iPhone has been shaped in different places. It will be especially interesting to hear how the iPhone has fared in countries where mobile media have been much more advanced than, say, the United States – for instance, in Japan, where mobile Internet was so much a part of both cell phone and Internet experience, and where the mobile has proved so protean and culturally significant. Here, the iPhone could be placed in at least two histories of new media adaptation – that of the histories of the mobile and its consumption, and that of the Internet.

As well as the critical vocabulary of cultural adaptation we might borrow from other areas of media, such as television and film, we also need to think about the concepts that are associated with technology. To understand the triple-play of adaptation in the case of the iPhone, for instance, we need to engage with the new lexicon of copying, modification, authorized and unauthorized use, versioning, and so on that comes from computer, Internet and digital cultures. What emerges here is a set of new considerations for understanding both the processes of adaptation, and the dynamics of culture that subtends them.

References

Apple. 2008. Accelerometer. Made to move. http://www.apple.com/iphone/features/accelerometer. html

Apple Insider. 2008. First iPhone 3G tear-down photos live from New Zealand launch. http://www. appleinsider.com/articles/08/07/10/first_iphone_3g_tear_down_photos_live_from_new_ zealand_launch.html

The Australian. 2008. Extras in iPhone queue. 26 August: 37.

Benkler, Y. 2006. *The wealth of networks: How social production transforms markets and freedom.* New Haven: Yale University Press.

Beschizza, R. 2007. Speaking freely: Unlocked, open source phones for weary iPhone hackers. 10 October. http://www.wired.com/gadgets/wireless/multimedia/2007/10/gallery_linux_phones

Boy Genius. 2008. Blackberry Bold vs iPhone 3G: Yeah, we definitely went there. 16 July. http://www.boygeniusreport.com/2008/07/16/blackberry-bold-vs-iphone-3g-yeah-we-definitely-went-there

Donner, J. 2009. Mobile media on low-cost handsets: The resiliency of text messaging among small enterprises in India (and beyond). In *Mobile technologies: From telecommunications to media,* ed. Gerard Goggin and Larissa Hjorth, 93–104. New York: Routledge.

Fortunati, L. 2006. User design and the democratization of the mobile phone. *First Monday* 7. http://firstmonday.org/issues/special11_9/fortunati/index.html

Fung, A. 2004. Coping, cloning, and copying: Hong Kong in the global television format business. In *Television across Asia: Television industries, programme formats and globalisation,* ed. A. Moran and M. Keane, 74–87. London: RoutledgeCurzon.

Geekipedia. 2007. iPhone. *Wired* magazine, 10 September. http://www.wired.com/culture/geekipedia/magazine/geekipedia/iphone

Goggin, G. 2006. *Cell phone culture: Mobile technology in everyday life.* London: Routledge.

Grossman, L. 2007. Invention of the year: The iPhone. http://www.time.com/time/specials/2007/article/0,28804,1677329_1678542,00.html

Helft, M., and J. Markoff. 2007. Google enters the wireless world. *New York Times,* 5 November. http://www.nytimes.com/2007/11/05/technology/05cnd-gphone.html?_r=1&ex=1352005200&en=d7a169e184415788&ei=5088&partner=rssnyt&emc=rss&oref=slogin

Hjorth, L. 2009. *Mobile media in the Asia Pacific: Gender and the art of being mobile.* London: Routledge.

Hutcheon, L. 2006. *A theory of adaptation.* London: Routledge.

Johnson, J. 2008. Is the iPhone the next Wii? *WiiNintendo,* 7 March. http://www.wiinintendo.net/2008/03/07/is-the-iphone-the-next-wii

Keane, M., A. Fung, and A. Moran. 2007. *Out of nowhere? New television formats and the East Asian cultural imagination.* Hong Kong: Hong Kong University Press.

Keane, M., and A. Moran. 2005. (Re)presenting local content: Program adaptation in Asia and the Pacific. *Media International Australia* 2005, no. 116: 88–99.

Law, P.-L., L. Fortunati, and S. Yang, eds. 2006. *New technologies in global societies.* Singapore: World Scientific.

Ling, R. 2008. *New tech, new ties: How mobile communication is reshaping social cohesion.* Cambridge, MA: MIT Press.

McCarthy, C. 2007. Live from New York, it's the iPhone launch! *Crave* blog. 29 June. http://news.cnet.com/8301-17938_105-9737706-1.html

Moran, A. 1998. *Copycat TV, globalization, programme formats and cultural identity.* Luton: University of Luton Press.

Moran, A., and M. Keane, eds. 2007. *Television across Asia: Television industries, programme formats and globalisation.* London: RoutledgeCurzon.

OpenMoko. 2008. Neo FreeRunner. 26 September. http://wiki.openmoko.org/wiki/Neo_FreeRunner

Rose, Frank. 2005. Battle for the soul of the MP3 Phone. *Wired* magazine. 13, no. 11. http://www.wired.com/wired/archive/13.11/phone.html.

Sawhney, H. 2009. Innovations at the edge: The impact of mobile technologies on the character of the Internet. In *Mobile technology: From telecommunications to media,* ed. Gerard Goggin and Larissa Hjorth, 105–17. New York: Routledge.

Turkle, S. 2007. Introduction: The things that matter. In *Evocative objects: Things we think with,* 1–11. Cambridge, MA: MIT Press.

Vogelstein, F. 2008. The untold story: How the iPhone blew up the wireless industry. *Wired* magazine. 16, no. 2. http://www.wired.com/gadgets/wireless/magazine/16-02/ff_iphone.

Wikipedia. 2008. 'Spirit level'. 27 September. http://en.wikipedia.org/w/index.php?title=Spirit_level&oldid=241333457

Zittrain, J. 2008. *The future of the Internet – and how to stop it.* New Haven, CT: Yale University Press; http://futureoftheinternet.org/download

'Pigeon-eyed readers': The adaptation and formation of a global Asian fashion magazine

Jinna Tay

Introduction

Fashion magazines are one of the most successfully 'glocalized' or adapted media in the world. Across newsstands in almost any city, a diverse range of fashion titles can be found littered among the urbanscape and, despite the perhaps confusing array of titles on offer, visitors are able to zero in on their weekly or monthly supply, or find an alternative that suits their taste. While the contents of magazines may be textually different around the globe, they conform to certain structural rules which allow their target readership to find them. From *I-D* to *Vogue and Another Magazine* to *Elle*, they will have a celebrity female form or face on the cover, with the issue title spread across in a big, bold font. In some, the subtitles on the bottom or side of the magazine promote the features and fashion spread, often in bright clear fonts. While the style and content of the magazines may differ, between the covers we can expect to find the latest on fashion photography, updates of consumer and/or beauty products, feature articles on celebrities and lifestyle, advertisements and editorials. The reproduction of these conventions is key to understanding how fashion is mediated, constructed and then stylized for different tribes and target groups by modern magazines. The differentiation of each fashion magazine and the targeting of their ideal readership are achieved stylistically, textually and aesthetically by adapting the content to the specific tastes and interests of the readers. Thus it is possible to see that the process of copying and adaptation are 'twin synergies' which have to be implemented in the creation of a new fashion magazine. To do so, this article examines a successful globally distributed Asian fashion magazine, *WestEast*, produced in Hong Kong, and looks at the strategies it has had to adopt to make it globally successful as well as successful within the Asian fashion markets.

Adaptation and copying

Copying and reproduction have been a systematic way in which human activities, business practices, knowledge and cultural forms have been translated globally. In the media area, for example, copying has enabled television ideas, programs and genres to be reproduced around the world. This has occurred through trade in formats (Moran 1998; Keane and Moran 2008), cross-genre fertilization (Turner 2005) and through the global diasporic media industries (Sinclair and Cunningham 2000). However, copying is only part of the process of marketing an established product into a new market. It is impossible to think about re-creating a product or a system in new conditions without first understanding which bits need to be replicated (copied) and which bits need to be changed (adapted) to suit the local conditions.

Thus adaptation can be seen here as the active process of modifying a specific set of operations, ideas or formula with a specific end-product and outcome in mind. Specifically with the production of fashion magazines, there is an active process of tailoring, selecting and shaping the vast quantity of global fashion information into a style tribe, a manageable product listing, a fashion zeitgeist with featured articles and fashion designers and labels to suit the taste of that specific niche readership. Thus this article understands adaptation and copying in fashion magazines to have a symbiotic relationship that governs the differentiated processes which go into establishing a new cultural product in a set environment. It is my argument that the more producers are aware of the local production conditions, their readership and the cultural response to the global/local system, the further they can shape the product so that it is more nuanced and responsive to its target readers. This business practice of adapting and transforming a wider global logic to local conditions has also become known as 'glocalization' (Robertson 1995), and fashion magazines are a useful location through which to observe this practice in action.

Studying fashion magazines

Even though fashion magazines have gradually been recognized as an important component of media studies and validated as scholarly objects pertinent to the formation of modern identities, much of the literature still regards the commercial aspect of magazines as a corrosive influence on the medium. While a study by Paul Jobling (1999) addresses this commercial aspect of magazine production in *Fashion Spreads*, it still does not take into account how magazines become localized. Therefore, to conceive of the process of adaptation as part cultural, part business acumen is relatively new.

The existing research on fashion magazines in the media and cultural studies area often questions the impact, production and readership of these magazines, and seeks to explain them within the framework of social or political relevance. As such, the literature can be divided into four general categories, although there is some overlap here. The first category centres on magazines and the development of the feminine public sphere and emancipation for women; here, fashion and women's magazines are integral in establishing a 'world of women' such that it allows the female readership to identify or produce a sense of its own resistance, subversion or dominance (Winship 1987; Ballaster 1991; Weiner 1999). The second category of writing investigates how magazines promote certain forms of knowledge formation, and ways of reading in particular, with niche readers such as teenage girls, middle-aged women, and so on (Hermes 1995; Skov 1995; McRobbie 1997). This also taps into the thesis of magazines as being 'windows to the world' for their readers and society, whereby they can see, read and be introduced to knowledge which perhaps is not readily accessible – such as pictures of the White House, or the lifestyle of celebrities. The third category focuses on readers as victims or hapless consumers, pointing to the negative effects of magazines – especially on younger readers such as teenagers (Sakamoto 1999).

The last category draws on the relationship that magazines have with the nation and the interface between the urbanscape and modernization; examples include discussions of post-war France (Weiner 1999), Japan (Sakamoto 1999) and Hong Kong (Fung 2002). Much of this literature explores how fashion magazines have performed the role of cultural intermediary and at times actively influenced the politics of a nation that is in transition. However, it is notable that within the study of fashion and women's magazines there is a significant gap in the literature on the business and commercial operations of magazines as cultural and media activities. Yet the commercial imperative is an important one within the magazine as it sets up the financial and productive framework, defining and limiting its creative capacity; it also determines how the material, textual and aesthetic aspects of the magazine are played out. The analysis on *WestEast* magazine that follows engages with the last two categories of analysis by taking on an examination of the adaptation process that allows these cultural negotiations to take place while also asking what forms of negotiations and identities are facilitated through *WestEast*'s success within Asia.

Syndication versus independent titles

Labelling a magazine as syndicated or independent indicates how that magazine is formed and owned. Fashion magazines such as *Cosmopolitan* have a good reputation for high circulation across many of the countries in which they are syndicated. A syndicated magazine would have an existing template of content, layout, font and typography to work from, and this is usually supplemented by a ready pool of photographers, editorial assistants and management expertise from which to draw. It would also most likely have an established distribution network, which comes with target readership profile and investment dollars. Retailers and distributors might already be familiar with its international title, and thus would be more open to stocking it on their shelves (Caperton 2003). To put it simply, syndicated publications refer to the industry practice of establishing a magazine in a new market based on an original title and format that has been largely successful with a cult (*Wallpaper*) or popular following (*Vogue*) elsewhere. Typically, it combines content produced by the original team with additional local content from the target city to inject a local feel and relevance into the new title. Syndication has been a successful means of international strategic business expansion for the magazine publishing industry (Caperton 2003). As a business model, it is similar to the practice of franchising. However, the business arrangements for each syndicated title may differ in each national context; some publishing houses take on sole publishing rights and others have co-publishers such as *Elle Singapore*, which is published by Mediacorp (Singapore) and Hachette Filipacchi (the original publishers). Also, the proportion of content that is shared between the 'parent' and the 'child' publication may range from 60–40 to 50–50.

The preference for syndicating fashion magazines arises out of a calculated business strategy to reduce risk by building on a known (international) title and brand such as *Vogue*, *Marie Claire* or *Cosmopolitan*. Alan Zie Yongder, the publisher of Hong Kong *Marie Claire*, *Esquire* and *Penthouse*, points out that what would normally take 20 years to establish can be done in one to two years through the use of international branding. However, what is also clear is that relying on an international brand is not enough to attract a local following – the magazine has to localize to succeed (Wilson 1995). The subject of fashion, in particular, relies on global savvy not only about the top couture houses in the world but also the local market stalls. The answer to 'what is fashion or fashionable' is still very much about what comes out of fashion cities such as London, New York, Paris and Milan. However, this is not to discount the influence of street fashion; rather, it recognizes that the large couture and production houses still maintain hegemonic control over the fashion industry. A large part of how that influence is established globally is through fashion magazines.

Independent publications are set up without any brand affiliations. In some cases within the Asian market they are individually owned but in others they are semi-government owned (Mediacorp and Singapore Press Holdings) or corporate conglomerates (*South China Morning Post*). In smaller cities such as Singapore and Hong Kong, their umbrella publishers like *The Straits Times* (Singapore Press Holdings) and *South China Morning Post*, respectively, have corporate structures that allow them to establish syndicated and independent titles with larger regional distribution channels. With the revenue stream in magazine business changing over the years, publishers rely mainly on advertisement, commercial artwork and sometimes subscriptions to deliver their profit margin (Lee 2004). So individually owned magazines have to compete against not just globally syndicated titles but also the bigger bucks of regional publishers. That said, it is telling that the top-selling magazines in countries like Singapore and Hong Kong are mainly the local independent magazines *Her World*, *Teenage*, *Citta Bella* and *Nuyou* (Singapore) and *Next Magazine*, *Yes Magazine* and *Jessica* (Hong Kong).

Yet, no matter how a magazine may be established, owned and run, larger stylistic conventions are copied to help readers to first recognize that it is a fashion magazine and, second, to adapt the content through the selection and identification of styles that are interesting to its target readers. This is where the process of adaptation kicks in. Imagine that you are the executive team commissioning a fashion title, or an editor/entrepreneur attempting to set up a fashion publication – how would you distinguish your magazine? What semiotic and stylistic devices, advertising and promotions would you use to attract and tempt your readers? What aspects from the massive databank of global fashion couture, designs, personalities and products would you offer to your readers? How would the fashion spectacle be tailored and selected to suit local tastes (which maybe conservative or traditional), and how would that selection of fashion images and information connect with the central feature stories and/or thematic advertisements? In this sense, the challenges associated with glocalizing materials offer us a way of examining adaptation in a different light – as a form of cultural adaptation, or the adaptation and transformation of cultural forms.

WestEast magazine

> I think I am selling a philosophy and a concept … Welcome to the premiere issue of *WestEast* [*WE*], the magazine that endeavours to bring together the best of two worlds. Like most of our readers, *WE* is fluent in more than one language and the product of more than one culture. (Kevin Lee, editor, *WestEast* magazine)

In 2001, Kevin Lee, a 20-something fashion journalist turned independent editor and publisher, launched his first magazine, *WestEast*, in Paris. As a fashion journalist in Paris for Taiwan *Fashion TV*, Lee decided to start a fashion magazine from Asia about Asians, selecting Hong Kong as his base of operation. As it later turned out, there were strategic marketing and cultural reasons behind Lee's decisions to base himself in Hong Kong. Lee's motivation for a new fashion magazine began in Paris when he noticed that Asian art, design and fashion industries seemed to lag behind global standards. His idea was to produce a high-profile fashion magazine that showcased a mix of current Asian culture, urban hotspots, and fashion and celebrities, as well as Western ones. He felt that this would contribute to the reverse cultural flow of fashion texts arriving from the West. Most importantly, this had never been done successfully in Asia (Lee 2004).

WestEast incorporates a high-end, at times avant-garde, style of fashion photography by renowned photographers, also attracting top models and celebrities such as Linda Evangelista, Devon Aoki and Kylie Minogue. The feature articles explore the emerging Asian 'indie' artistic scene, Asian popular culture, travels in Asia, fashion and architecture in Asia, Asian traditions and emerging subcultures. While they included timely Western icons such as Peter Jackson in New Zealand during the screening of *Lord of the Rings*, the early issues of *WestEast* concentrated on Asian celebrities and icons such as Japanese pop princess Ayumi Hamasaki, Hong Kong actress Gigi Leung and then emerging Chinese star Zhang Ziyi. Beyond the A List fashion photography and celebrities, *WestEast* also profiled popular artists such as Michael Lau, a creator of collectible vinyl figurines and Lynn Ban, a Chinese New Yorker and well-known collector of vintage fashion. In doing so, *WestEast* can be seen as aggregating these global 'Asian' cultural activities into one cohesive text, culturally bridging and fostering a two-way exchange in the flow of popular texts between East and West.

To begin with, Lee's pedagogic intention was for *WestEast* to serve as a platform to disseminate and champion fashion-related Asian styles, features and celebrities, tapping into the growing global interest in twenty-first-century Asia. As a fashion text, *WestEast* hybridizes Western and Eastern ideas; this fusion runs through the entire magazine from its structure and featured subjects to its fluent visual style. Fusion of West and East serves as a structuring device for the overall execution and production of the magazine and as a way of targeting its upwardly mobile middle-class readership. *WestEast*'s success regionally and globally (the magazine is now in its seventh year and has expanded into a new title, *WestEast Men*) is significant both as a case study for understanding the cultural barriers it has negotiated as well as the textual and marketing strategies it had to employ as a means of adapting the fashion magazine genre into a particular context.

Within a year of bursting onto the scene, *WestEast* had managed to establish good connections and a solid reputation on the local and global media circuit, as evidenced by the pro bono work of iconic photographers and celebrities, and the growing contributions from a global network of feature editors from the East (Shanghai, Tokyo) to the West (New York, Milan, Paris, Berlin and London). Furthermore, the publication received plenty of media recognition and accolades from the *South China Morning Post*, a leading newspaper for Hong Kong and the region. To top off that first year's work, *WestEast* won several international magazine awards, including Excellence in Magazines (June 2003), Best Three Independent Style Magazines (May 2003) and Excellence in Magazine Design (June 2003) (www.westeastmag.com). This series of developments ensured the magazine plenty of publicity among local and global readers.

Bringing about East and West?

WestEast's first issue in 2001 (top left cover in Figure 1) showcased cover girl Devon Aoki's mixed parentage as a physical epitome of the East–West cultural hybridity that

WestEast espouses. Her reflection of this duality is a signposting of the fusion and cultural diversity that *WestEast* claims to instil and provoke (even if it is a little too obvious). Aoki herself muses:

> I think it's really important to have Asian girls on the catwalk – it changes the face of fashion, diversifies it, stretches the boundaries and creates more possibilities for what can be beautiful. It's amazing to see a black girl on the cover of American Vogue, and it will be just as amazing to see an Asian girl on the cover! (Belverio 2001, 39)

These basic arguments for multicultural representation have been widely circulated by scholars such as Shohat and Stam since 1994. As a further demonstration of his point, Lee explains that the logo of *WestEast* – two circles (see Figure 1) – is a play on the duality of West and East where: 'The solid circle represents a Western full stop, with the outline O (*juhao*) reflecting an Asian one' (Lee 2001, 13). Thus the logo presents a neat, coherent and compact representation of this duality and exchange. Lee states: '*WestEast*'s philosophy is one of sharing' (13), a promotion of intercultural exchange. Though the *WestEast* logo is a compact metonym of the duality and exchange of ideas of West and East, such simple idealistic representations are often sites of disjunctures that require deeper textual excavation, especially when they gloss over complex identifications of East and West. So we look to the fashion text for other clues to how it is dealing with this issue.

If *WestEast* aims to be a transmission point and cultural bridge between East and West, then it is important to consider whether *WestEast*, as a self-proclaimed Asian magazine, showcases Asianness distinctively. How are East and the West showcased in its pages? In subsequent issues, *WestEast* carried features and original, artistically shot photographs of celebrities – actors, models, fashion designers – dominated by Asian figures such as Maggie Quigley, a popular Hong Kong celebrity from a non-Chinese background, Ai Tominagi, an international Japanese model, and Josie Ho, a Hong Kong actress and daughter of a Macau tycoon. Most of the features were inclined towards 'rediscovering' a wider definition of China, from the showcase of Chinese ballet dancers to Western 'fashion designers in Asia' (meaning China) and architectural innovation taking place along the Great Wall of China or Beijing's disappearing *Siheyuan* (courtyard houses).

Figure 1.   *WestEast* magazine covers.

It also focused on the emerging new Chinese film directors and *Budaixi*, a popular and traditional form of Taiwanese puppet theatre.

However, what stands out among all these features is a lengthy investigative feature on the underground punk rock scene at (in)famous Su Village, Beijing, written by Shaway Yeh. It is probably not coincidental that Yeh, a Taiwanese, is also the guest editor of this issue. She lives in New York, edits and writes for international and Asian publications such as *V Magazine* and *Dutch*, and is in the process of moving to Beijing. A detour through this feature will illustrate the process of modernization in China and how it has been adapted into the main narrative of fashion and subcultures in China through the prism of fashion magazines.

Entitled 'I am a middle-aged groupie' (Yeh 2001), this is a feature with vivid black and white photo-shoots over five pages. Yeh describes the scene of a community of young musicians who are extremely financially challenged and living in rundown huts on the outskirts of Beijing. There are no creature comforts such as toilets or shower facilities, but they choose to live this way as a statement of rebellion against the affluence and expectations of their increasingly middle-class families. They look to Western notions of freedom and democracy, carrying the idealism of Woodstock to be replicated in China. Their angst and disenchantment are inspired by figures such as Kurt Cobain and Limp Bizkit. This story of self-marginalized youth is an interesting departure from the norm, highlighting the alternative underbelly of China, an aspect that is unsanctioned and often goes uncovered in the present stratum of news. Yeh wrote bemusedly that, at the end of the last night, the boys all shot their fists in the air and screamed 'Freedom!' in a style learnt from their favourite Pasolini movie, *Salo*.

This story is resonant in its rawness and contains a duality that comes from the subjects adopting advanced avant-garde, alternative postures, tagged as 'suffering for their art' while the general country is developing and modernizing rapidly. However, the earth-shattering processes of modernization often produce equally robust disparate subjects seeking alternative identities, a disavowal of mainstream success and a refusal to conform. This social phenomenon is not unique to present-day China. Indeed, it seems to be a by-product and a characteristic of modernity. However, that this exemplar of a rising subculture in China comes out of *WestEast* demonstrates how *WestEast* has situated itself as an 'indie' fashion magazine that uncovers hitherto unexplored phenomena in Asia, offering cutting-edge narratives and hip fashion visuals – thus positioning itself as a cool, authoritative and original fashion magazine from Asia. At a deeper level, this story also posits an identity that is distinctively Chinese without any of the usual orientalist Chinese stereotypes or mainstream representations such as Hong Kong celebrities.

The magazine's obvious capacity to speak fluently in the global fashion visual language, and its determination to showcase an alternative popular aesthetics, combined with its links to global talents and circulation, demonstrates that it has managed to find its target readership and further adapt its unusual content to the confines of a fashion magazine. *WestEast* conforms to central structural conventions of the fashion magazine, such as the glossy high-quality paper, high-end fashion photography and the singular figure of a celebrity on the front cover illustrating the theme of that particular issue – for example *Sex*, *New China*, *Generasian*, *Movement*, presenting a clear strong clean visual statement of the magazine and its theme. The visual layout is similar to other high-end niche fashion magazines such as *I-D* or *Another Magazine*, indicating the kind of readership to which it is marketed: those who take fashion and subcultures seriously and engage with them.

The readership profile from 2003 provided by *WestEast* gives us a more concrete indication of the readership breakdown. It is telling that *WestEast* attracts the readership of the AA Group: they are aged from 20 to 40, with an almost equal male-to-female ratio, and three-quarters of the readers are single. A large proportion of these readers (72%) have been educated in 'Western' countries. Only 56% live in the 'West', while most have an annual income level of US$30,000–50,000. There was no mention of the sexual preference of the readership. In terms of occupation, the highest number of readers worked in the banking sector, then finance and publishing. The profile of the readers of *WestEast* shows that, while they have equal numbers based in the East as well as West, in effect this readership comprises a well-paid, urban, youth- and culture-oriented cosmopolitan group that is interested in the rising Asian culture and its contributions to global fashion, design and subcultural affiliations. Just as the data profile offers a glimpse into *WestEast*'s target market, it is important to locate Hong Kong as the mediascape where these production issues are absorbed and the decisions of adaptation and copying made.

Cultural production and the 'pigeon-eyed' phenomenon

In terms of its production space, *WestEast* is located in a small office with its editorial team based in a side street in central Hong Kong. The office consists of five people, including Lee and Guo. Lee has a personal assistant and two other art and graphic designers. The rest of the editorial correspondents and writers are to be found in New York (Glenn Belverio and Lin Ting Ting), London (Lee Kuanting), Milan (Wagner Raimondi), Tokyo (Michiru Shimano), and so on. Lee travels widely and frequently to meet up with the rest of his team to collaborate on stories and visuals as he finds 'his inspiration everywhere' (Lee 2004).

Guo and Lee state that most of the photographers, stylists and contributors work pro bono as *WestEast* cannot afford to pay them. Citing the first anniversary issue of *Sex* with Kylie Minogue on its cover as an example, Guo explained how Minogue agreed to pose for them free of charge after their photographer (an acquaintance of Minogue's) approached her and gave her a copy of *WestEast*. Lee believed they were very lucky and said most collaborators had contributed out of love for the project. He believes that the incentive for the collaborators is that they can be as experimental as they want to be.

As *WestEast*'s main source of income is derived from advertising, Guo states that the magazine's strategy of charging advertisers the 'local rate' for a global circulation has proved to be a success. Advertisers are pleased that their ads have a global presence and this directly translates to the taglines of a global commodity – 'London, Paris, New York' – and the absence of a local address for many of the advertised brands. By its first anniversary issue, *WestEast* was commissioned to make in-house ad copy for advertisers such as Vertu, and since then several other advertisers.

The interviews I conducted with Kevin Lee and the marketing manager Jeannie Guo offered insight into their understanding of the global fashion industry in the context of cultural consumption in Asia, particularly Hong Kong. Lee reflected that while living in Paris he observed that the Chinese seemed to lag 'right behind everyone else' (2004) in the field of fashion, art or design. His contention was that the general Asian mentality was to copy directly, with no innovation. Consequently, Asia was not producing any good-quality fashion magazines that represented its ideas, arts or culture. With this in mind, he initially toured France and America, speaking to more than 100 people about his idea. In the final stages, he went to Asia to round up some talent, but the response was one of disbelief and ridicule. Lee stated: 'Asians don't believe in themselves. You can't put the blame on others when you can't believe in yourself … They [Singaporean photographer] laughed at me.' To illustrate his point further, he stated that a month before *WestEast* was scheduled to go to print, an Asian investor pulled out, preferring to invest in a Chinese version of *Dutch*. From his initial research, he realized that for the magazine to succeed in Asia, it would have to be successful in Western fashion capitals first. He calls it the 'pigeon-eyed' phenomenon.

Both Jeannie Guo and Kevin Lee, in separate interviews, described this cultural phenomenon that Hong Kongers term 'pigeon-eyed'. It roughly translates to the act of Asians who look down on Western cultures or foreign products, or who consistently believe that they are much better than anything locals can produce. This was the reason why *WestEast* took this strategic route to sell the magazine. Lee stated that:

> In my strategy, it is West and East mix. In France they respect culture and countries with cultural background. I wanted to modernize but it could not be too obvious, too Chinese, so I naturally mixed both sides, sold it internationally. With good sales overseas, I imported it back to Asia. So we had to make a circle, do it well overseas before taking it back.

Lee believes that this particular strategy of arriving in Asia via Europe has established *WestEast*'s credibility in the fashion industry. The pigeon-eyed attitude is obviously commonplace enough for it to be attributed a direct Cantonese slang. However, Lee contends that it is by no means restricted to Hong Kong but is pervasive throughout Asia, most notably in the area of goods and services. Fashion labels from Italy, France, the United States or the United Kingdom are perceived to be more prestigious and fashionable than local ones. Even successful High Street brands from Japan or Hong Kong are regarded as of lower status than labels like H&M from the United Kingdom, just as European appliances and European cars are perceived to be more luxurious than the Asian-made ones.

This has deep historical roots. Cultural theorist and anthropologist Ong Aihwa notes that, back in the 1960s, British education was considered the best and Asia was frequently constructed as 'failed replicas of the modern West' (Ong 1996, 60). This second-tier mentality is what Lee seeks to address. His argument is that the global fashion world should see what contemporary Chinese culture has to offer. Ironically, the challenge to revise this image of Asia must first be brokered by using his understanding of Asian pigeon-eyes to attract the approval of the European fashion industry and readership in order to pave the way for its Asian acceptance.

Hong Kong as semiotic port

At the beginning of this paper, one of the questions posed enquired about the context of Hong Kong in relation to the fashion texts. By intentionally choosing Hong Kong as the site of *WestEast*'s production, what are the factors that drew Lee to it? After analysing the textual intentions and discourses in *WestEast*, how do they relate to the wider urbanscape and issues in Hong Kong?

At its most denotative level, Hong Kong exists through its popular entertainment industry – for example, its television and film celebrities, Canto pop and media talents. It is especially recognized as such in East and Southeast Asia; Hong Kong's media personalities are its face, the personification of its talent pool, media capabilities, professional production and the centre of Chinese media production. Lo (2005) notes that Hong Kong has played a major role in representing Chineseness and Chinese popular culture to the world and, despite the current downturn in film production (it is currently tenth in the world) (MapsofWorld 2005), it is still recognized as such. Lo maintains that Hong Kong is a 'prolific production centre of Chinese diaspora culture and one of the most important platforms for Chinese–Western cultural mediation' (Lo 2005, 2). Media and cultural theorist Eric Ma (2005) argues that local history was deliberately ignored in formal education under the British, and history was never a discursive tool disposed towards myths of national construction as they are frequently utilized by other nations (153–4). Within that space, popular media have stepped in to foster a sense of 'community and history' that relates more to the 'opportunities of the present and a vision of the future' (Ma 2005, 154) than to the past. In fact, rather than a national discourse, Hong Kong's lack of a distinctive nationalist structure under British/Chinese colonialism has produced a cultural identity that has tended to negate its Chineseness in order to appear 'less parochial and more modern' (Lo 2005, 3). As such, it occupies a peculiar space, an Asian culture that has managed to thrive on cultural rather than national narratives (Ma 2005, 153).

Situated between political indeterminacy and an energetic economy that was closely tied to the globalized market system prior to 1997, it is to the 'local culture' of the everyday to which the Hong Kong people turn, to make sense of their everyday lives (Lo 2005, 27). Hong Kong's media have always taken, as their point of cultural reference, narratives predisposed to either local diasporic experiences, which were largely Chinese, or to a globalizing narrative. Lise Skov's (2004, 225–8) in-depth study of the history of the Hong Kong fashion industry observes that Hong Kong became the functioning export-based garment manufacturer for global fashion houses in Europe and America. It became a global wholesale market for accessing material, accessories and skilled tailoring from Shanghai migrants. Perhaps Hong Kong's lack of political outlet and its consistent locale as a bridging medium have also served its media productions and talent pool well. Hong Kong, as the metaphorical bridge, has served the region for British interests, for global capital and for regional exchanges.

Thus it is no accident that Lee has chosen to base *WestEast* in Hong Kong. In terms of publicity, *WestEast* has a strong foothold in the local press, with Lee himself seen as a celebrity with social standing as the editor of *WestEast* magazine (HKTDC 2004). He is seen as 'one of the eligible guys' hanging out with famous local actors such as Edison Chan. By associating himself with media personalities, Lee is able to promote *WestEast* but also take advantage of the credibility of *WestEast* to extend his industry network. The Asian celebrities in turn appear in features and photo-shoots in *WestEast*, grant interviews and pose for avant-garde fashion photography (pro bono). Taking advantage of Hong Kong's strong media culture and celebrity-tabloid hungry readers, *WestEast* is well positioned to keep up with the news but also make news. Lee also mentions that, as a media-rich production centre, Hong Kong has a good talent pool of artistic and graphic designers he can easily tap into.

Consequently, it is possible to understand that Hong Kong can in fact, be seen as the 'semiotic port' of Asian culture, especially over the last 20 years or so (Lo 2005, 8). Historically, it has successfully re-created and expanded the repertoires of 'authentic' historical Chinese mythic and fictional identities to which all forms of 'Chinese' cultural values have been ascribed. In place of a slumbering closed-door China, Hong Kong has been the popular and successful disseminator of this pan-regional diasporic sense of Chineseness. Thus it is not too far off the mark to suggest that the text and function of *WestEast* parallel Hong Kong's semiosis, and because it is able to become a semiotic port so effectively it consequently facilitates the production of interesting culturally rich and diverse fashion texts which span the topics of art, dance, architecture, design, history and sexuality (to name a few). It is also important to note the crucial role played by Hong Kong culture in mediating and translating Chinese culture on the one hand, and copying, modifying and adapting Western culture on the other.

Conclusion

Within the discourse of establishing new fashion titles, even at its conceptual stage, adaptation has to play a part determining what the magazine should look like, who it can target and how that could be done. Adaptation in this context involves the process of tailoring fashion information which is both global and local into the right mixture so that it presents both a credible knowledge of global fashion trends and an understanding of how the local scene may interpret and indigenize these trends to suit the local conditions.

These discussions respond to the fact that adaptation as a general process is widely practised but often overlooked as a component of media production and analysis. In glocalizing fashion magazines, where producing a global feel with local flavour is essential, the process of adapting and synthesizing allows the fashion information to be tailored to the local reader; or vice versa, *WestEast*'s synthesizing of local stories for global fashion consumption becomes a way of distinguishing and of locating the magazine in the global magazine marketplace.

The process of adaptation is easy to overlook because it seems like a 'natural response' – a logical, strategic process to manage the shifting environment. Yet, while adaptation is recognizably a large part of the marketing strategies of indigenizing and glocalizing, it is larger than both because it is not just a top-down or bottom-up approach but is in fact constitutive of the mode of transformation. Thus it is engaged and called upon in any scene of active transformation.

References

Ballaster, R. 1991. *Women's worlds: Ideology, femininity and the women's magazine*. London: Macmillan.

Belverio, G. 2001. Devon and earth. In *WestEast*, Issue 1, Winter, 36–9.

Caperton, K. 2003. Forecast '03: Go East. *Folio: The Magazine for Magazine Management* 32: 33.

Fung, A. 2002. Women's magazines: Construction of identities and cultural consumption in Hong Kong. *Consumption, Markets and Culture* 5, no. 4: 321–36.

Guo, J. 2004. Interview with author, Hong Kong.

Hermes, J. 1995. *Reading women's magazines*. Cambridge: Polity Press.

HKTDC. 2004. Hong Kong Young Fashion Designers' Contest 2005 promotes new talents and Hong Kong's fashion industry. Hong Kong Trade Development Council. http://info.hktdc.com/tdcnews/0410/04101601.htm (accessed 15 October 2008).

Jobling, P. 1999. *Fashion spreads: Word and image in fashion photography since 1980*. Oxford: Berg.

Keane, M., and A. Moran. 2008. Television's new engines. *Television & New Media* 9, no. 2: 155–69.

Lee, K. 2001. Editor's word, Issue I, Winter.

Lee, K. 2004. Interview with author. Hong Kong.

Lo, K.-C. 2005. *Chinese face/off: The transnational popular culture of Hong Kong*. Urbana: University of Illinois Press.

Ma, E.K.-W. 2005. Re-advertising Hong Kong: Nostalgia industry and popular history. In *Asian Media Studies*, ed. J. N. Erni and S. K. Chua, 136–58. Malden, MA: Blackwell.

MapsofWorld. *c*.2005. World map showing top 10 producing feature films. http://www.mapsofworld.com/world-top-ten/world-top-ten-feature-film-production-map.html (accessed 15 October 2008).

McRobbie, A. 1997. MORE! New sexualities in girls' and women's magazines. In *Back to reality? Social experience and cultural studies*, ed. A. McRobbie, 190–209. Manchester: Manchester University Press.

Moran, A. 1998. *Copy cat TV: Globalisation, program formats and cultural identity*. Luton: University of Luton Press.

Ong, A. 1996. Anthropology, China and modernities: The geopolitics of cultural knowledge. In *Future of anthropological knowledge*, ed. H.L. Moore, 90–100. London: Routledge.

Robertson, R. 1995. Glocalization: Time–space and homogeneity–heterogeneity. In *Global modernities*, ed. M. Featherstone, S. Lash, and R. Robertson, 25–44. London: Sage.

Sakamoto, K. 1999. Reading Japanese women's magazines. *Media, Culture & Society* 21, no. 2: 173–93.

Sinclair, J., and S. Cunningham. 2000. Go with the flow: Diasporas and the media. *Television & New Media* 1, no. 9: 11–31.

Skov, L. 1995. Environmentalism seen through Japanese women's magazines. In *Women, media and consumption in Japan*, ed. L. Skov and B. Moeran. Honolulu: University of Hawaii Press.

———. 2004. Fashion shows, fashion flows: The Asia Pacific meets in Hong Kong. In *Rogue flows: Trans-Asian cultural traffic*, ed. K. Iwabuchi, S. Muecke, and M. Thomas, 221–46. Hong Kong: Hong Kong University Press.

Turner, G. 2005. Cultural identity, soap narrative, and reality TV. *Television & New Media* 6, no. 4: 415–22.

Weiner, S. 1999. Two modernities: From *Elle to Mademoiselle*. Women's magazines in post-war France. *Contemporary European History* 8, no. 3: 395–409.

Wilson, C. 1995. Clones of Western magazines thrive in Asia. *International Herald Tribune*, 20 February. http://ww.iht.com/articles/1995/02/20/magcon.php (accessed 21 February 2001).

Winship, J. 1987. *Inside women's magazines*. New York: Pandora Press.

Yeh, S. 2001. I am a middle-aged groupie. In *WestEast*, Issue 1, Winter, 144.

Craic in a box: Commodifying and exporting the Irish pub

Bill Grantham

Craic/crack and the Irish pub concept

In a recent Irish comedy film *Dick Dickman PI* (O'Neill 2008), an incompetent private eye investigates a wave of disappearances of Irish traditional musicians. In a culminating police swoop on a Russian gang at Rosslare Harbour, County Wexford, a packing crate on the quay is prised open to reveal a pub band in full swing. 'They're smuggling crack!' exclaims the PI.

With apologies for explaining (and thereby spoiling) the joke: the word 'crack' in Hiberno-Irish is a noun and verb that can mean 'entertaining chat, sport; to have fun' (Dolan 2002–05). It's an English/Scottish dialect word that originally referred mainly to good conversation and most likely entered Ireland through Scottish settlers in the north of the country, and somewhere along the line, at least by the 1960s (*Connacht Sentinel* 1968, 3), was Gaelicized into *craic* (pronounced the same way) with added connotations of 'ribaldry and divilment' (Ó Muirithe 1996, 81). It has become the quintessential term among the Irish for having a good time. And the fact that *craic*/crack can be stolen and smuggled suggests another aspect of the concept: it is a *commodity* that can be bought, sold, branded, manufactured, packaged, put in a box and, above all, exported (McGovern 2003). Indeed, that is exactly what is happening – all over the world:

> People in Ireland have crack all the time. Especially in pubs. And they're not doing anything illegal. That's because they're having '*craic*', Gaelic for fun. When you open an Irish pub, you are in fact, importing *craic* from Ireland. You will want to hear your customers spread the word that there's 'mighty *craic* to be had altogether' in your Irish Pub.
>
> How can you ensure that this will happen? We advise abiding by the eight Critical Success Factors to ensure that the *craic* will be ninety in your pub. (Diageo-Guinness USA 2008)

Cultural note: 'the *craic*/crack was ninety' refers colloquially to a particularly elevated level of *craic*/crack. The phrase is found in a song, 'The Crack was Ninety in the Isle of Man' by Barney Rushe, who played it for the Irish musician Christy Moore as early as 1968. Rushe later owned an Irish pub in Germany (Moore 2008; Daly 2008).

On its Irish Pub Concept website, Diageo-Guinness USA, the American subsidiary of Diageo plc, a London-based alcohol conglomerate, provides a video to answer the pressing question: 'What's the *Craic*?' It's a canny simultaneous appeal to 'tradition' and modernity – Irish-sounding music with an electric bass line, images of Irish hills and carved wooden pub fixtures blended with hortatory marketing slogans in faux-Celtic script: ENJOY THE BENEFITS OF A FRANCHISE WITHOUT THE DOWNSIDE ... 98% SUCCESS RATE IN US ... TYPICALLY UNDER 3 YEARS RETURN ON INVESTMENT ... THE US IS AN UNTAPPED MARKET ... WORLD'S MUST SUCCESSFUL PUB CONCEPT. As a definition of *craic*/crack, there's no ribaldry or divilment about it at all at all. Instead, you'll need an upfront investment of around $1.5 million for an 'average' pub, although Diageo will provide you with a pro forma business plan to get you started for just $150.

The 'eight Critical Success Factors' for an Irish pub (it turns out that there are only seven), according to Diageo, are as follows:

1.   *'Location, location, location'*

How, Diageo asks, will you know if you are 'in the right market for Irish target customers?' Diageo provides a 'site location finder' (a map) to assist you: the 'hottest opportunities in the US' are to be found where there is 'low IPC penetration to date' (an 'IPC' is an 'Irish Pub Concept'), an 'excellent consumer fit' and where Diageo recommends targeting 'metro areas'. Most highly recommended locations include California, Washington, Nevada, Montana, Colorado, Texas, Minnesota, Florida, Georgia and the region north of Boston. With the exception of Boston, these are not the places generally associated with the Irish population of the United States – originally Massachusetts, Illinois, New York and New Jersey, although there has been latter-day migration of Irish Americans to sunbelt states such as California, Arizona and Florida (Almeida 2006, 550, 569).

2.   *'Irish staff + your Irish pub = an authentic Irish experience'*

While it is possible, Diageo says, to 'recreate the feel of a true Irish pub without Irish staff, we don't recommend it'. Curiously, one place where Irish pubs increasingly lack 'Irish staff' is Ireland. In recent years, large numbers of immigrants from Eastern Europe, but also sub-Saharan Africa and Asia, have arrived in Ireland. Non-native Irish bar staff are routine. In increasingly multicultural Dublin, one city-centre pub is owned by a Korean. A local pub in a fishing village in County Clare might be staffed by Latvians, Lithuanians and Poles (Helm 2006; Lavery 2006; Lynch 2006). But for Diageo, 'No Irish pub' – for which, read 'IPC' – is 'complete without the friendly warmth, humor and advice of a true Irish bartender'. The 'true Irish bartenders' on the Diageo site are from Central Casting – white, of course, and often red-haired. Unmentioned in this conjuring of this 'authentic Irish experience' is that the practice of choosing staff of one nationality over another is also illegal in the United States, as elsewhere. The Civil Rights Act of 1964 states that it is 'unlawful' for an employer to:

> fail or refuse to hire or to discharge any individual, or otherwise to discriminate against any individual with respect to his compensation, terms, conditions, or privileges of employment, because of such individual's race, color, religion, sex, or national origin. (Civil Rights Act 1964, 2000e–2(a)(1))

And why is this 'authenticity' – apparently exclusive of the growing numbers of Irish of, say, Nigerian or Chinese origin – so vital? According to Diageo, 'authenticity' leads to 'sales per square foot in current authentic IPC pubs [which] are exceeding the U.S. average by a factor of two. Additionally, the IPC beverage dominated product mix has led to attractive profit margins in these pubs.' This is the secret part of the *craic*/crack equation: the 'Authentic Irish Experience' (however illegally obtained) = Enhanced Profits.

3. *'Stout and a song'*

Stout, naturally – the black ale is what Guinness sells, although these days it sells more of it in Nigeria than in Ireland (Murphy 2008). As for the 'song': it turns out that there is no need to kidnap Irish musicians and put them in crates for foreign export. Instead, in addition to live music, pub proprietors are urged to attend to the 'background tunes' that 'are considered so essential to the atmosphere' that 'operators' need to 'use sophisticated sound systems', among which Diageo recommends the DMX Profusion X, which 'bases music selections according to time/energy of day so they can be harmonized with the mood and pace of the clientele'. Helpfully, these 'systems' also 'include computerized updates over the Internet'. DMX Music, in fact, is the 'exclusive supplier[s] of music systems to the Irish Pub Concept in the United States', according to Diageo. DMX 'uses consumer lifestyle trends, market segmentation, and branding expertise to design compelling music experiences' (DMX Inc. 2008). It is not clear whether these 'consumer lifestyle trends' include the fact that in 'actual' – as opposed to IPC-virtual – Irish America, the 'Irish traditional music community' is 'thriving', 'myriad weekly music sessions are held in large and small cities' and 'record labels are producing recordings by Irish and Irish-American musicians at an unprecedented rate' (Miller 2006, 415). It does not appear relevant to the mission of designing compelling – profitable – 'musical experiences'.

4. *'True Irish food and drink'*

> Can you dine on traditional Irish fare without a Guinness in front of you? We think not. Authentic Irish beverages, particularly great Irish draught beers like Guinness Stout and Harp Lager, are at the heart of the Irish Pub Concept and can be large profit generators.

However, according to Diageo, 'a true Irish pub uses food to drive beverage sales, having recognized an important cultural difference among Americans', the difference apparently being that 'while Irish and Europeans socialize around drinks, U.S. society socializes around food'. (This would be news to the French, Spanish and Italians, but also to the contemporary Irish, with their Michelin-starred restaurants and modish gastropubs.) Guinness has accordingly come up with an 'Irish Pub Food Manual' for the American publican to use in order to provide 'traditional Irish fare' – such as Guinness Beef Crostini, whose Hiberno-Italian concept combines a French *baguette* with beef, Guinness and 'Cashel Bleu [*sic*] Cheese Sauce' – 'prep time 2 minutes', 'cook time 8 minutes'.

Cultural note: Cashel Blue is a delicious blue cheese from County Tipperary. Its long tradition stretches all the way back to 1984, when it was first developed by a farming couple. Guinness's Gallicized 'Bleu' creates ethnographic confusion, which is as nothing compared to the gastronomic infamy of making a sauce out of the cheese and slopping it on top of a beef-Guinness stew: the furthest in this direction its makers will go is to suggest grilling a little on top of a steak (J & L Grubb Ltd. 2008).

Another 'Irish' menu abandons the pretence of Irishry altogether, offering the likes of muffins, croissants, bagels and 'traditional' espresso drinks (GGD Global 2008). At the Mother Egans, Fado and BD Riley pubs in Austin, Texas – 'you can't get more any Irish than BD Riley' – the menus include salmon with sesame seeds and rice, roasted vegetables with hoisin sauce, stuffed jalapenos and fajitas (Rice 2001, E1). Ironically, these offerings are closer to what actual Irish people are actually eating and drinking these days, the same as everyone else. But they're not 'authentic' in the sense, say, that Irish stew is 'authentic'.

Of course, the principal goal of Diageo-Guinness USA is for these Irish pubs to import not only *craic* but also Guinness and the company's other 'premium drinks' from its 'outstanding collection of beverage alcohol brands across spirits, wine and beer categories'. Draught Guinness was created only in 1959 and has been so constantly refined since then that in taste, texture and temperature it is completely different from the original eighteenth-century form of the drink; in any case, the Guinness 'Extra Stout' on which Draught is loosely based was traditionally outsold by a light beer, porter, which in 1904 accounted for three-quarters of the company's income (Molloy 2002, 65). Whatever Guinness is selling today, it is hardly 'traditional'. In addition to Draught Guinness, a well-stocked pub will offer such Irish brands as Harp Lager, first sold ('launched', in Diageo-speak) in 1960, Smithwick's Ale, 'launched' in 1710, Baileys Irish Cream ('the result of a remarkable recipe combining the finest ingredients with just the right amount of Celtic charm'), 'launched' in 1974 as the result of the efforts of a 'committee of senior managers in Gilbeys of Ireland [that] had the idea for a uniquely Irish drink reflecting Ireland's heritage and unparalleled agricultural and distilling traditions' and Bushmills Whiskey, which has a genuine claim on tradition, having been distilled in County Antrim since it was 'launched' in 1784 (Diageo 2008). Also, Hennessy cognac, 'launched' in 1765 by émigré Irish distillers in France, and apparently beloved by North Korea's leader Kim Jong Il (Blitzer 2003) and a host of Diageo booze brands suggestive of other borrowed traditions: Red Stripe beer, Moët & Chandon champagne, Tanqueray and Gordon's gins, Captain Morgan Rum, José Cuervo tequila, Smirnoff vodka and Bell's, Black & White and Haig scotch whiskies, among many.

5.   *'Irish design and build'*

How do you get your authentic Irish pub to *feel* Irish? You start with an Irish name, the choice of which is 'extremely important' and which 'tells a lot about your pub'. You might begin with a 'traditional city name, such as Limerick's' (note that there appears to be no pub in County Limerick actually called 'Limerick's' [*Golden Pages* 2008]), or else look 'to create the illusion of history with a name like O'Keeffe & Sons'. Diageo provides a list of more than 400 'Irish' names to help out: Aherne's, Altamont Gardens, Ambrose Magillicuddy's, *An Casúr* [The Hammer], *An Cruiscín Lán* [The Full Jug)] *An Currach* [The Currach-boat], Anderson, Auld Dubliner, Baggott Inn, Ballybunnion, Ballygoan ... and so on.

After the name comes the look. According to Diageo: 'Most of the traditional Irish pubs in the U.S. are built from the ground up by specialty firms ... [that] recreate the warmth of an Irish pub using the same materials and designs that have stood for centuries in Ireland.' One of these, the Irish Pub Company, based in an industrial estate outside Dublin, offers five styles of pub: Country Cottage, Brewery, Gaelic, Traditional and Victorian Dublin, each of which is provided with its own back story. For instance, Country Cottage:

> The design for this pub derives from the traditional stone-built style of an Irish country cottage, many of which were so welcoming and hospitable that over the years they simply became accepted as the local pub.
>
> Originally, to become a publican, the master of the house would erect a sign bearing his name over a brightly painted front door, and arrange an enticing display of drinks in the front window to attract thirsty customers.
>
> Thereafter, the publican would open up his house to neighbours and friends for cheerful hours of drinking, story-telling, conversation and music. Inside, a warm atmosphere would be

created by rough white-washed plaster walls, timber beams, stone floors, and dressers stacked with tankards and brightly colored crockery.

Customers would sit upon wooden benches or in settle-seats and a large open fire – complete with grate, bellows and a suspended cast-iron kettle – would provide a hub for the convivial conversation of the evening. (Irish Pub Company 2008)

Another designer, the Ól Irish Pub Company (the *ól* refers to an Irish word for drink), like Diageo, appeals to tradition stacked against modern cool:

Ireland and Irish culture is a big pulling factor of the Irish pub concept. Thanks to the world-wide success an [*sic*] culture rich history crammed with writers and poets such as James Joyce, Seamus Heaney and William Yeats to name a few and more recently the musical talents of U2, The Corrs and The Cranberry's [*sic*]. Add to this the well-renowned friendliness of the Irish and you quickly understand why people of all nationalities choose an Irish pub as their watering hole. (Ól Irish Pub Company 2008)

A competitor, GGD Global, 'the market leader in Irish Pub Designs for over three decades', has come up with a 'concept' to address a common problem: insufficient space for your Irish pub. The 'Snug Irish Corner' is an 'ingenious, low-cost, high-revenue outlet' that 'fuses two world-beating concepts – contemporary Celtic café culture and the traditional appeal of the Irish pub – to create an all-day food and drink offering in a stylish and intimate Irish setting' (GGD Global 2008). With its 'remarkably small investment' and 'high day and evening revenues' which are 'guaranteed', the Snug Irish Corner is an 'even more exciting opportunity' than the IPC. Diageo, too, has its 'Snug' concept, with illustrated installations at sports arenas, as well as 'portable pubs' and 'pouring stations' that are 'movable to maximize traffic patterns and visibility opportunities' and 'perfect for sports and concert venues, hotel banquets, stations, airports or patios', all combining 'Irish authenticity' with claimed low investment requirements and high returns.

A full pub can be planned and built in 26–31 weeks (Diageo 2008). A design company will visit your site, make a proposal 'comprised of detailed authentic renderings of how the pub will appear', provide project management, 'manufacture all elements' including woodwork, furniture, lights, glasswork, metalwork, bric-a-brac and the like, then transport it to you (clearing customs for you on the way) and 'install' the pub (Irish Pub Company 2008). The 'bric-a-brac' is considered particularly important since, as Diageo points out: 'Framed posters, signs, jugs and old bicycles … could help create a comfortable Irish Pub environment' and 'make your pub look authentic'. However, acknowledging that this stuff may take 'many years to collect', Diageo instead recommends using a 'Guinness Approved Supplier' that allows 'any pub [to] order authentic Irish bric-a-brac', each 'based on an original piece of classic advertising'. Once ordered, the *craic* arrives from Ireland in a box, ready to become 'authentic' in its new home. No wonder that, as one writer puts it, 'Ireland, as much of the world knows it, was invented in 1991', the year the Irish Pub Company was formed (Kelley 2006).

6. *'Good management'*

As Diageo points out, a 'solid management team that has a love and understanding of the Irish Pub Concept will add greatly to your success'. To this end, Diageo provides an 'Operations Management Training Course in Dublin, home of Guinness in order to learn more about managing your Irish Pub'. Another player, the Celtic Dragon Pub Company, offers 10-day staff training courses with written exams, because for a pub operator with a knowledgeable staff, 'the more you learn, the more you earn' (Celtic Dragon Pub Company 2008).

7.   *'Positive environment'*

Despite the quest for authenticity, Diageo admits, based on 'commissioned research', that the top factors keeping customers in their pubs are 'air quality' and 'cleanliness of bar and toilets', as well as 'music', 'staff' and 'food offering'. This is the one place in the IPC pitch where women – practically invisible graphically on the IPC website – emerge as crucial ingredients of the 'experience'. 'Did you know', Diageo asks:

> that when a mixed group of men and women visit your Irish Pub, the women will decide if the group at large will stay in your pub for any length of time? Women do not like Irish pubs that are too cramped, smoky, noisy, dirty or a place where the staff is not friendly and helpful.

Women? Clean toilets? Hoisin sauce? Just what kind of authentic Irish pub are we talking about, anyway?

The Irish pub in history and memory

During the Celtic Revival of the late nineteenth and early twentieth centuries CE, there grew an antiquarian fondness for using ancient texts as exemplars of ostensibly contemporary Irish traits, particularly those that might distinguish the native Irish from their British – especially English – neighbours. Thus, in the Instructions of King Cormac Mac Airt, attributed to a semi-legendary Irish high king and collected in the fourteenth century CE, the 'duties of a chief and of an ale-house' are enumerated:

> Good behaviour around a good chief,
> Lights to lamps
> Exerting oneself for the company
> A proper settlement of seats
> Liberality of dispensers,
> A nimble hand at distributing
> Attentive service
> Music in moderation
> Short story-telling
> A joyous countenance
> Welcome to guests
> Silence during recitals
> Harmonious choruses. (Meyer 1909)

This combination of pub ideals – liberal ale, fast and gracious service, music, talk and hospitality (if not silence during recitals) – could have inspired the IPC copywriter. The licensed trade in Ireland is fond of this fantasy. It has connected itself to the 'most honoured ranks in Celtic society, that of *briugu*, or hospitaller, who was only worthy of the status if he had 'a never-dry cauldron, a dwelling on a public road and a welcome to every face'. The *briugu* trope appears in a history of the Irish pub commissioned by a trade association, and in the association's submissions to a parliamentary committee reviewing alcohol policy (Molloy 2002, 1; O'Donoghue 2006). The truth that gets in the way of this good story is that the *briugu* was not a Celtic proto-publican but a *nouveau riche* social climber, in the words of an eighth century CE legal text, a 'wealthy man of non-noble birth [who] could acquire high rank through displaying ... hospitality and generosity' (Kelly 1988, 36). Foundation myths abound. Brewing was alleged to have been introduced to Ireland by one Partholón, a descendant of Noah, who arrived in Ireland 300 years after the Great Flood (Macalister 1938–53, §38). Attacks on alcohol abuse were deflected by reference to the Christian water–wine miracle at Cana (O'Donoghue 2006).

These myths seek to locate in the contemporary Irish pub a tradition that lies at the heart of the definition of Irishness. They even attempt to trump alternative loci of 'authenticity'. Thus, for one modern chronicler of the pub:

> Some may speak of Irish Christianity or literature as being the great contribution of our wet little island to the greater world beyond these shores. The true connoisseur of Irish culture knows, however, that these are pale shadows compared to the worldwide influence of our public houses and alcoholic beverages. (Blake and Pritchard 1997)

Even allowing for a spot of mock-heroic ironizing, the basic sentiment here is pure mush.

This connection between the Irish pub and the construction of Irishness had been taken on board by the internal tourist industry before the launch of the globalized Irish pub in the 1990s, emerging as 'the site for a form of cultural tourism in which a socially constructed conception of Irish people is consumed in a setting that also allows for tourist escape, hedonism and exploration' (McGovern 2003, 84).

However, this movement – internally and externally – is taking place at a time when the Irish pub in Ireland appears to be in decline. Changing social practices, longer commutes, smoking bans and drink-driving law enforcement have combined in rural areas to drive many pubs out of business (Jordan 2008, A01). For every pub that has closed, it is claimed, 'a dozen more' have been denatured through being 'hurriedly 'modernized' with salt and pepper canisters on every table, homogeneous fitted furniture and giant plasma screens blasting out sport and music' (Fennell and Bunbury 2008, 10). Regardless of its cultural image, the 'Irish pub has probably passed its peak' (Malcolm 1998, 50).

The truth is that the historical moment of the 'Irish pub' – as opposed to the mere pub situated in Ireland or foreign tavern serving expatriates – has been relatively brief. There have always been lots of them, which in one sense means that they were and are nothing special – in 2002, there were some 12,500 pubs, mainly family-owned businesses, on the island of Ireland serving a population of some 5.6 million, or roughly one pub for every 450 people (Molloy 2002, 1; CSO 2008; NISRA 2008). In the mid- to late seventeenth century CE, there were an estimated 1500 taverns in Dublin alone, serving a population of less than 60,000, or one for every 40 people (Molloy 2002, 29; Craig 1952, 21). In 1908, there were more than 17,000 pubs for a population of around 4.4 million, or one for every 255 people (in Tralee, County Kerry, the number at that time was one in 80) (Ferriter 2005, 57). When the population is adjusted for the principal customers of the pubs, adult males, the concentration is even greater: in the latter part of the nineteenth century CE, there was one pub for every 40 in Cork city and one for every 35 in Waterford; nearly one-fifth of the houses in Limerick were pubs (Malcolm 1998, 51). They were as much mundane features of everyday life as special sites of cultural personality.

The reputations of the pubs and taverns were also nothing special. In sixteenth century CE Galway, the taverns were said to be dirty with 'neither sitting place, cloth, dish or any other service' (Malcolm 1998, 56). The English chronicler Barnabe Rich (a bitter witness, having been marooned in Ireland and disappointed in life, career and money, but probably on to something all the same) railed in 1610 against the Dublin taverns that were, he said, 'the very nurseries of drunkennesse, of all manner of idlenesse, or whoredom, and many other vile abominations' (Gilbert 1978, i.151–2). Outside the cities with their English (or English-aligned) populations, the ordinary native Irish 'did not have pubs and seem to have drunk alcohol irregularly and often in conjunction with special occasions'. This began to change in the late seventeenth century CE, when consumption of spirits began to increase rapidly, and government sought to boost revenues by taxing alcohol and licensing the premises on which it was sold, while at the same time limiting the availability of drink at outdoor events. Alcohol consumption was driven indoors, subject to the oversight and control of national laws, courts, excise enforcement and the newly constituted Irish police. Out of these efforts, the 'Irish pub' emerged from the 1850s onwards (Malcolm 1998).

However, some at least kept the grim allure of the Elizabethan and Jacobean taverns of Galway and Dublin. In nineteenth century CE Ulster, it was said that 50,000 people consumed 17,000 gallons of diethyl ether – made by combining grain alcohol with sulphuric acid – per year, much of it in shebeens (Connell 1996, 90). The English novelist Anthony Trollope – not as unsympathetic as many of his contemporaries to the Irish and Irish life – wrote in the 1840s of an illegal shebeen in Mohill, County Leitrim, that was 'a resort of the wicked, the desperate and the drunken', offering 'freedom from decent restraint' and 'the power of inebriety at a cheap rate' (Trollope 1906, 129). The County Mayo shebeen in *The Playboy of the Western World* (1907) is 'very rough and untidy' with 'a sort of counter' filled with 'many bottles and jugs' and 'empty barrels', with an open fireplace and a single table (Synge 1968, 107). A century later, in Frank McCourt's memoir *Angela's Ashes*, the pubs of Limerick are dark, sad places, where the family's scant funds are drunk away by a feckless father (McCourt 1996). (Which, apparently, if improbably, did not prevent at least one pub in Limerick from renaming itself in honour of the book [Foster 2001, 174].)

The association of the drinking-place with drunkenness is a natural one, albeit not prominent in the marketing of the IPC. By 1738, it was being claimed that Ireland had a 'relish and a love' of the 'beastly vice' of drinking, while others connected alcohol abuse with public disorder and crime, including murder (Garnham 1996, 181). The final words of executed criminals in eighteenth century CE Ireland – formulaic accounts of repentance following decline into decadence and criminality – contain frequent references to excessive drinking on the path to ultimate disaster (Kelly 2001). The rates of arrest for drunkenness in Dublin in 1871 were more than eight times those of London (O'Brien 1982, 188). The prevalence of drunkenness led to a mobilization against it, most notably in the temperance crusade launched in 1838 by Father Theobald Mathew that reportedly enlisted hundreds of thousands to the cause of total abstinence (*Catholic Encyclopedia* 2008). The Irish Association for the Prevention of Intemperance, founded in 1878, campaigned for pubs to be closed on Sundays and for opening hours to be reduced (O'Brien 1982, 187). The Pioneer Total Abstinence Association, launched in 1898, 'went on to become the largest Catholic lay movement in twentieth century Ireland and, indeed, as a percentage of the population, one of the largest movements of its kind in the world'. As late as 1970, there were nearly 2700 Pioneer centres in Ireland (Ferriter 2005, 57, 594).

This association of the Irish with heavy drinking permeated abroad, and heavy drinking became an index of Irishness, both within and without Irish communities. In Irish America: 'Public drinking, especially among males, was central to becoming Irish-American; the more one drank, and was seen to drink, the more Irish one became.' From the 'fact of Irish drinking' emerged 'the stereotype of the Irish drunk, violent or comic'. Substantially more Irish alcoholics were admitted to American hospitals and institutions than those of other nationalities or origins (Kenny 2000, 200–1). In English cities in the 1950s and 1960s, Irish immigrants accounted for a disproportionate number of arrests for drunkenness and were viewed by the native population as being 'on the whole a people addicted to drinking to excess' (Delaney 2007, 181–3). The bars, saloons and pubs of the Irish diaspora were similarly identified by natives with the unfavoured Irish and therefore embraced by the dispersed Irish themselves. In American cities, saloons lay at the centre of Irish political organization, and were therefore distrusted by those who lay outside the Irish political machines (Moynihan 2007, 482; Quinn 2007, 675). In England,

'Irish' pubs – usually simply 'English' pubs that attracted large numbers of migrants – often were a 'substitute for home' with 'conviviality in an ostensibly familiar environment' and permission to 'relax without fear of offending the "natives"' (Delaney 2007, 174). These were inwardly regarding places, unwelcoming to outsiders and offering the experience of 'Irishness' to the Irish alone.

When did everyone else become Irish?

The IPC insists that 'revenue at authentic Irish Pubs exceeds typical Irish bar sales by as much as 400% in some markets'. Which is to say that sales in 'authentic' Irish pubs – those organized upon IPC lines – easily outstrip 'typical' Irish bars – those that have grown in Irish and Irish-American populations. In other words, the key to Irish 'authenticity' in the IPC universe is to differentiate and distance the pub from 'actual' Irishness, apart from the red-headed bar staff. This effort is doubtless helped by positive tourist experiences from visiting Ireland, film and television depictions, and writerly accounts of Irish pub life from James Joyce (in *Ulysses*, the convivial lunch at Davy Byrne's, more than the encounter with the ranting anti-Semite Citizen in Barney Kiernan's), Flann O'Brien's *At Swim-Two-Birds* to various literary memoirs (Joyce 1960; O'Brien 1967; Fallon 1998, 149–57; Cronin 1999, 9–10). However, it is also part of the Irish pub consumer wanting to belong, a more complex construction of the 'Kiss Me, I'm Irish' impulse displayed on St Patrick's Day T-shirts. But what 'Irish' is this? If you accept that the contemporary Irish are as likely to be espresso drinkers, crostini-munchers, hoisin-saucers and digitized-music-experience consumers as anyone else in the globalized world economy, then the offerings of the IPC are not much different from those of you local Starbucks, KFC or Armani: simply more hues in the multi-brand palate of transnational consumerism. And there's no particular reason why the intangibles of consumption – the time, the place, your mood, the other people present – should not make any one IPC bar into a perfectly good place to drink and pass time. It may be that, just as American mobsters have apparently begun to talk like the Corleones and the Sopranos, the IPC and other 'Irish' branding efforts may have the effect of constituting a new 'real' Irishness. But in physical, contemporary Ireland, it's one that the Korean pub-owner, the Latvian server, the Pakistani bar-manager, the Chinese kitchen worker, the Nigerian Guinness drinker and the rest of the 'unreal' Irish will have difficulty appreciating. For a people so long disfavoured by others, it shocks to see them turned into antagonists of difference, often for the benefit of those who once disfavoured them, or their descendants.

References

Almeida, L.D. 2007. Irish America, 1940–2000. In *Making the Irish American*, ed. J.J. Lee and M.R. Casey, 548–573. New York: New York University Press.

Blake, L., and D. Prichard. 1997. *The Irish pub*. County Wicklow: Bray.

Blitzer, W. 2003. North Korean leader loves Hennessey [*sic*], Bond movies. 8 January. http://www. cnn.com/2003/US/01/08/wbr.kim.jong.il (accessed 20 October 2008).

Catholic Encyclopedia. [1917] 2008. Theobald Mathew. http://www.newadvent.org/cathen/10047a. htm (accessed 25 October 2008).

Celtic Dragon Pub Company. 2008. Celtic Dragon Pub Company – staff training. http://www.celticdragonpubco.com/cel_stafftrain.html (accessed 25 October 2008).

Central Statistics Office Ireland (CSO). 2008. CSO – statistics: Population 1901–2006. http://www.cso.ie/statistics/Population1901-2006.htm (accessed 25 October 2008).

Civil Rights Act 1964. 42 U.S.C. 2000e.

Connacht Sentinel. 1968. Teach Furbo ag oscailt anocht ceol agus craic [Furbo House open tonight music and *craic*]. *Connacht Sentinel*, 30 July (Galway).

Connell, K.H. 1996. Ether-drinking in Ulster. In *Irish Peasant society: Four historical essays*. 87–111. Blackrock, County Dublin: Irish Academic Press.

Craig, M. 1952. *Dublin 1660–1860: A social and architectural history*. Dublin: Hodges Figgis.

Cronin, A. 1999. *Dead as doornails*. Dublin: Lilliput Press.

Daly, P. 2008. Paul Daly: A musical biography. http://www.create.moonfruit.com/#/pauldaly/4516778520 (accessed 20 October 2008).

Delaney, E. 2007. *The Irish in post-war Britain*. Oxford: Oxford University Press.

Diageo. 2008. Diageo. http://www.diageo.com/en-row/homepage.htm (accessed 20 October 2008).

Diageo-Guinness USA. 2008. Irish pub concept. http://www.irishpubconcept.com/homepage.asp (accessed 20 October 2008).

DMX Inc. 2008. DMX – make your brand unforgettable/full sensory branding experience. http://www.dmx.com (accessed 25 October 2008).

Dolan, T.P. 2002–05. A Hiberno-English archive. http://www.hiberno-english.com (accessed 12 October 2008).

Fallon, B. 1998. *An age of innocence: Irish culture 1930–1960*. Dublin: Gill & Macmillan.

Fennell, J., and T. Bunbury. 2008. *The Irish pub*. New York: Thames & Hudson.

Ferriter, D. 2005. *The transformation of Ireland*. Woodstock, NY: Overlook Press.

Foster, R.F. 2001. *The Irish story: Telling tales and making it up in Ireland*. London: Allen Lane/Penguin.

Garnham, N. 1996. *The courts, crime and the criminal law in Ireland 1692–1760*. Blackrock, County Dublin: Irish Academic Press.

GGD Global. 2008. GDD Global – Irish pub design and management. http://www.ggdglobal.com/index.html (accessed 20 October 2008).

Gilbert, J.T. 1978. *A history of the City of Dublin* [1854–59]. Dublin: Gill & Macmillan.

Golden Pages. 2008. *Golden Pages* – product, business and service search directory for Ireland. http://www.goldenpages.ie/displayhome.ds (accessed 25 October 2008).

Helm, J. 2006. Irish opportunity attracts Poles. *BBC News*, 31 January. http://news.bbc.co.uk/1/hi/world/europe/4661716.stm (accessed 25 October 2008).

Irish Pub Company. 2008. The Irish Pub Company, Irish pubs abroad, interior designs, consultants and architects, fixtures and fittings. http://www.irishpubcompany.com (accessed 25 October 2008).

J & L Grubb Ltd. 2008. Cashel Blue Irish Farmhouse Cheese. http://www.cashelblue.com/index.htm (accessed 20 October 2008).

Jordan, M. 2008. In affluent New Ireland, rural pubs are so yesterday. *Washington Post*, 25 April: A01.

Joyce, J. [1922] 1960. *Ulysses*. London: The Bodley Head.

Kelley, A. 2006. Ireland's 'crack' habit. *Slate*, 16 March. http://www.slate.com/id/2137893 (accessed 20 October 2008).

Kelly, F. 1988. *A guide to early Irish law*. Dublin: Institute for Advanced Studies.

Kelly, J. 2001. *Gallows speeches from eighteenth-century Ireland*. Dublin: Four Courts Press.

Kenny, K. 2000. *The American Irish: A history*. Harlow: Longman.

Lavery, B. 2006. Dublin Journal; Now, the barkeeps may come from the ends of the earth. *The New York Times*, 16 May. http://www.nytimes.com/2006/05/16/world/europe/16dublin.html?scp=1&sq=Jae+Hyuk+Lee&st=nyt (accessed 25 October 2008).

Lynch, T. 2006. When Latvian eyes are smiling. *The New York Times*, 17 March. http://www.nytimes.com/2006/03/17/opinion/17lynch.html?n=Top/News/World/Countries%20and%20Territories/Ireland (accessed 25 October 2008).

Macalister, R., trans. 1938–53. *Lebor Gabála Érenn* [The book of the taking of Ireland]. Irish Texts Society. http://www.ancienttexts.org/library/celtic/ctexts/lebor2.html (accessed 20 October 2008).

Malcolm, E. 1998. The rise of the pub: A study in the disciplining of popular culture. In *Irish Popular Culture 1650–1850*, ed. J.S. Donnelly and K.A. Miller, 50–77. Dublin: Irish Academic Press.

McCourt, F. 1996. *Angela's ashes: A memoir*. New York: Simon & Schuster.

McGovern, M. 2003. The cracked pint glass of the servant: The Irish pub, Irish identity and the tourist eye. In *Irish tourism: Image, culture and identity*, ed. M. Cronin and B. O'Connor, 83–103. Clevedon: Channel View Publications.

Meyer, K., trans. 1909. *Teocosca Cormaic* [The instructions of King Cormac Mac Airt]. Todd Lecture Series. http://www.ancienttexts.org/library/celtic/ctexts/cormac3.html (accessed 20 October 2008).

Miller, R.S. 2006. Irish traditional music in the United States. In *Making the Irish American*, ed. J.J. Lee and M.R. Casey, 411–416. New York: New York University Press.

Molloy, C. 2002. *The story of the Irish pub*. Dublin: Liffey Press.

Moore, C. 2008. Christy Moore: The official website. http://www.christymoore.com/lyrics_tabs_detail.php?id=16 (accessed 20 October 2008).

Moynihan, D.P. 2007. The Irish (1963, 1970). In *Making the Irish American*, ed. J.J. Lee and M.R. Casey, 475–525. New York: New York University Press.

Murphy, K. 2008. Column one. *Los Angeles Times*, 12 July. http://articles.latimes.com/2008/jul/12/world/fg-guinness12 (accessed 25 October 2008).

Northern Ireland Statistics and Research Agency (NISRA). 2008. *Northern Ireland population, 1901–2007*. http://www.nisra.gov.uk/demography/default.asp3.htm (accessed 25 October 2008).

Ó Muirithe, D. 1996. *The words we use*. Blackrock, County Dublin: Four Courts Press.

O'Brien, F. [1939] 1967. *At Swim-Two-Birds*. Harmondsworth: Penguin.

O'Brien, J.V. 1982. *'Dear, dirty Dublin': A city in distress, 1899–1916*. Berkeley: University of California Press.

O'Donoghue, S. 2006. Submission to Joint Committee on Arts, Sport, Tourism, Community, Rural and Gaeltacht Affairs. *Seanad Éireann*, 9 March. http://debates.oireachtas.ie/DDebate.aspx?F=TOJ20060329.xml&Ex=All&Page=2 (accessed 25 October 2008).

O'Neill, B. 2008. *Dick Dickman PI*. Ireland: 93 minutes.

Ól Irish Pub Company. 2008. Old Irish Pub Building Company – bar and restaurant design. http://www.olirishpubs.com (accessed 20 October 2008).

Quinn, P. 2007. Looking for Jimmy. In *Making the Irish American*, ed. J.J. Lee and M.R. Casey, 663–679. New York: New York University Press.

Rice, D. 2001. Pub grub grows up. *Austin American-Statesman*, 14 March (Austin TX): E1.

Synge, J.M. [1907] 1968. *The playboy of the Western world. Plays*, ed. A. Saddlemyer. Oxford: Oxford University Press.

Trollope, A. [1847] 1906. *The Macdermots of Ballycloran*. London: John Lane.

Afterword: Albert and Michael's recombinant DNA

Toby Miller

> Ready for a free, fun, no-hassle virtual makeover? The Makeover-o-Matic virtual makeover game lets you try on virtual hairstyles, makeup and accessories with your own photo or a model photo. Find your best online virtual makeover look and style using the latest beauty products, without the risk! Select from hundreds of hairstyles, cosmetic colors and accessories in the privacy of your own home. Blend, highlight, mix and match to create your new online look. What are you waiting for? Go ahead and get beautiful! (http://beauty.ivillage.com/0,,9jlxfdd5,00.html)

> Neo-liberalism is not Adam Smith; neo-liberalism is not market society, neo-liberalism is not the Gulag on the insidious scale of capitalism ... but ... taking the formal principles of a market economy and ... projecting them on to a general act of government. (Foucault 2008, 131)

Our editors have produced a remarkably stimulating and plural group of papers on a vital question for cultural studies today: whether we should regard culture as principally a collective or individualistic category (not all the papers were available to me when I wrote this piece – I received approximately two-thirds of the announced contents). Two huge forces play this issue out in ways that are referenced again and again in these contributions, albeit in a mostly implicit way: cultural difference and individual neo-liberalism. The first is reaching its apotheosis in the greatest collective endeavour of human history: twenty-first-century immigration and its consequences for cultural adaptation. The other is the greatest attempt to reshape individuals in human history via government programs of ideologization that focus on the freedom to choose via consumption. The two models have a complex relationship.

Immigration

Cultural adaptation has become a norm because of the collective experience of difference. Of the approximately 200 sovereign states in the world, over 160 are culturally hetero-geneous, and they comprise 5000 ethnic groups.[1] Between 10% and 20% of the world's population currently belong to a racial/linguistic minority in their country of residence. Nine hundred million people affiliate with groups that suffer systematic discrimination. Perhaps three-quarters of the world system sees politically active minorities, and there are more than 200 movements for self-determination in nearly 100 states. Even Anglo-Celts have been subdivided by cultural difference, as a consequence of both peaceful and violent action, and a revisionist historiography that emphasizes the millennial migration of Celts from the steppes; Roman colonization; invading Angles, Saxons, Jutes, Frisians, and Normans; attacking Scandinavians; trading Indians, Chinese, Irish, Lombards and Hansa; and refugee Europeans and Africans.

The number of refugees and asylum-seekers at the start of the twenty-first century was 21.5 million – three times the Figure 20 years earlier. The International Organization for Migration estimates that global migration increased from 75 million to 150 million people between 1965 and 2000, and the United Nations states that 2% of all people spent 2001 outside their country of birth, more than at any other moment in history. Migration has doubled since the 1970s, and the European Union has seen arrivals from beyond its borders grow by 75% in the last quarter-century. Five key zones of immigration configure today's world – North America, Europe, the Western Pacific, the Southern Cone and the Persian Gulf – along with five key categories: international refugees, internally displaced people, voluntary migrants, the enslaved and the smuggled.

No major recipient of migrants has ratified the United Nations' International Convention on the Protection of the Rights of All Migrant Workers and Members of their Families, even though they benefit economically and culturally from these arrivals. Opinion polling suggests sizeable majorities across the globe believe their national ways of life are threatened by global flows of people and things. In other words, their collective cultures are under threat. At the same time, they feel unable to control their individual destinies. In other words, their neo-liberal capacities are under threat. Majorities around the world oppose immigration.

Immigrants and their cultures have long been *the* limit-case for loyalty, as per Ruth the Moabite in the Jewish Bible/Old Testament. Such figures are both perilous for the sovereign-state (where does their fealty lie?) and symbolically essential (as the only citizens who make a deliberate decision to swear allegiance to an otherwise mythic social contract). Hence the regressive nationalism that greets them, in such forms as the belligerence of the United States, the anti-immigrant stance of Western Europe, or the crackdown on minorities in Eastern Europe, Asia and the Arab world. The populist corollary is often violent – race riots in 30 British cities in the 1980s; pogroms against Roma and migrant workers in Germany in the 1990s and Spain in 2000; migrant-worker and youth struggles in France in 1990 and 2005 – on it goes. The two most important sites of migration from the Third World to the First – Turkey and Mexico – see state and vigilante violence alongside corporate embrace in host countries, and donor nations increasingly recognizing the hybrid experience of daily life. This grand collective project of cultural adaptation, however accidental, voluntary, coercive, planned or casual, has made cultural recombination a fact of everyday life for both host and donor nations. But there is an alternative story contending with this narrative and its solid numbers – a fantastical account of personhood called neo-liberalism that does not have a research paradigm because it brings facts into being through the application of its theories rather than dealing with the world as it finds it.

Neo-liberalism

If the world of immigration and its multicultural legacy are at the core of cultural studies, then neo-liberalism is at the core of creative industries. The grand paradox of neo-liberalism is its passion for intervention in the name of non-intervention – pleading for investments in human capital at the same time as deriding social engineering; calling for the generation of more and more markets by the state while insisting on fewer and fewer democratic controls; and hailing freedom as a natural basis for life that can only function with the heavy hand of policing by government to administer property relations. Its lust for market conduct extends beyond such matters to a passion for comprehending and opining on everything, from birth rates to divorce, from suicide to abortion, from performance-enhancing drugs to altruism. Nothing can be left outside the market, and nothing left to the chance that market relations may falter without massive policing (Foucault 2008).

From this flows an entire host of methodologically critical matters in the growing distinction between cultural studies and creative industries: a focus on collective struggle over meaning versus the rush to the discourse of creativity and the subordination of politics; the refusal of neo-liberalism versus its embrace; and the methods of political economy (studying material conflicts) versus narcissography (watching TV or playing games with one's children and friends).

Albert and Michael's collection paves the way for us to consider anew what neo-liberalism was in the light of the cultural *frottage* caused by immigration. I use the past tense to describe neo-liberalism because the world's descent into the disasters of deregulation and offshoring has forced neo-liberalism's prelates, from Beijing to the Bourse, to rethink their dismissal of alternative norms (hint: Keynesianism). Dominant in world thought for three decades, neo-liberalism was nothing less arrogant than 'a whole way of being and thinking', an attempt to create 'an enterprise society' through the pretence that the latter is a natural (but never achieved) state of affairs, even as competition was imposed as a framework of regulating everyday life in the most subtly comprehensive statism imaginable (Foucault 2008, 218, 147, 145).

Neo-liberalism understood people exclusively through the precepts of selfishness. It exercised power on people by governing them through market imperatives, so that they could be made ratiocinative liberal actors with their inner creativity unlocked in an endless mutual adaptation with the environment. The market became a privileged 'interface of government and the individual' (Foucault 2008, 253). At the same time, the notion of consumption was turned on its head. Everyone was creative, no one was simply a spectator, and we are all manufacturing pleasure when we witness activities we have paid to watch. Internally divided – but happily so – each person was 'a consumer on the one hand, but ... also a producer' (Foucault 2008, 226). This is perhaps best exemplified in the media, a key aspect of Albert and Michael's collection.

Media

A television survey by *The Economist* in 1994 remarked that cultural politics is always so localized in its first and last instances that the 'electronic bonds' of exported drama are 'threadbare'.[2] Part of the talent of the cultural commodity is that it leads a lengthy career and can be retrained to suit new circumstances. To quote Liberace: 'If I play Tchaikovsky I play his melodies and skip his spiritual struggles ... I have to know just how many notes my audience will stand for' (Hall and Whannell 1965, 70). Culture is simultaneously the key to international textual trade and one of its limiting factors. Ethics, affect, custom and other forms of knowledge both enable and restrict commodification. In the 1960s, Disney television in Australia consisted of rebroadcast US programs; by the 1990s, it went through a superficial localization, via young Australian presenters. And General Motors, which owned Australia's General Motors Holden, translated its 'hot dogs, baseball, apple pie, and Chevrolet' jingle into 'meat pies, football, kangaroos, and Holden cars' for the Australian market. Hollywood recently won over a key segment of the Indian market with a localized *Who Wants to be a Millionaire?* This can be read as an indication of the paradigmatic nature of the national in an era of global companies, or as the requirement to reference the local in a form that is obliged to do something with cultural-economic meeting-grounds. In the end, the sale is always local. Of course, producers are the most powerful cultural adaptors: Hollywood got away with using the music, set and title of *Who Wants to be a Millionaire?* in *Slumdog Millionaire* without paying a cent, because of a rights-exploitation clause in the initial format sale. That is the reality of recombinant media culture.

Yet parts of cultural and media studies, especially the church of creativity, hew to a fetish of the individual – narcissography under the heavy sign of neo-liberalism. Irresistibly enchanted by a seeming grassroots cornucopia, struck by the digital sublime, many First World cybertarian technophiles attribute magical properties to today's communications and cultural technologies, which are said to obliterate geography, sovereignty and hierarchy in an alchemy of truth and beauty. A deregulated, individuated media world supposedly makes consumers into producers, frees the disabled from confinement, encourages new subjectivities, rewards intellect and competitiveness, links people across cultures, and allows billions of flowers to bloom in a post-political parthenon. In this Marxist/Godardian wet dream, people fish, film, fuck and finance from morning to midnight. The mass scale of the culture industries is overrun by consumer-led production, and wounds caused by the division of labour from the industrial age are bathed in the balm of Internet love. 'Everyone is a publisher' thanks to the Internet and its emblematic incarnation in social networks (Jenkins and Hartley 2008). These fantasies are fuelled and sometimes created by multinational marketers only too keen to stoke the fires of aesthetic and adaptive desire, as when Apple advertises 'Do-It-Yourself Parts for iBook' (apple.com/se/support/ibook/diy). *Time* exemplified this sovereignty of consumption in choosing its 2006 'Person of the Year'. The magazine famously announced the winner as: 'You. You control the Information Age. Welcome to your world' (Grossman 2006).

This apparent transformation is actually one more moment in an oscillation we have experienced routinely over the past century. During that period, each media innovation has offered people more of what they never knew they needed commercially, at the same time as it has promised new possibilities democratically. Today's touching cybertarian faith that individuals can control their destinies through the Internet, and folksy 'prosumers' can overpower big media thanks to their home-grown videos, is the latest version.

Academic cybertarians maintain that the new media provide a populist apparatus that subverts patriarchy, capitalism and other forms of oppression. All this is supposedly evident to scholars and pundits from their perusal of social media, conventions, web pages and discussion groups, or by watching their children in front of computers. Virginia Postrel wrote a 1999 *Wall Street Journal* op-ed in which she welcomed this Pollyannaish tendency within cultural and media studies as 'deeply threatening to traditional leftist views of commerce ... lending support to the corporate enemy and even training graduate students who wind up doing market research'. At such moments, we can say that what Terry Eagleton (1982) sardonically named The Reader's Liberation Movement is in the house. It can hardly be a surprise, then, to find Bob McChesney (2007) lamenting that contemporary media studies is 'regarded by the pooh-bahs in history, political science, and sociology as having roughly the same intellectual merit as ... driver education'. Or that the *Village Voice* dubs us 'the ultimate capitulation to the MTV mind' (Vincent 2000). Even Stuart Hall recently avowed: 'I really cannot read another cultural studies analysis of Madonna or *The Sopranos*' (MacCabe 2008, 14).

So let's be a wee bit more grounded here. Between 1980 and 1998, annual world exchange of electronic culture grew from US$95 billion to US$388 billion. In 2003, these areas accounted for 2.3% of Gross Domestic Product across Europe, to the tune of €654 billion – more than real estate or food and drink, and equal to chemicals, plastics and rubber. The Intellectual Property Association estimates that copyright and patents are worth US$360 billion a year to the United States, putting them ahead of aerospace, automobiles and agriculture in monetary value. Global information technology's yearly revenue is US$1.3 trillion. PricewaterhouseCoopers predicts 10% annual growth. And the cultural/copyright sector employs 12% of the US workforce, up from 5% a century ago (Miller 2009; Collins 2008). This is the underlying reality behind adaptive culture – its placement in, and impact on, the core of the world economy.

Take YouTube, putatively the acme of consumer-led cultural adaptation. But consider the thousands of contracts the firm has signed with mainstream media, and the introduction of Video Identification, a surveillance device for blocking copyrighted materials by tracking each uploaded frame. It spies on users and discloses their Internet protocols, aliases and tastes to corporations and permits these companies to block or allow reuse depending on their marketing and surveillance needs of the moment. The software was developed with those great alternatives to mainstream media dominance, Disney and Time Warner. Hundreds of companies have signed up in its first year. Sales of *Monty Python* DVDs on Amazon.com increased by 1000% since they became part of the system ('YouTube rolls out filtering tools 2007; Kiss 2008; for fine overviews of surveillance and social networks, see Andrejevic 2007; Cohen 2008). This is a cybertarian dream of individuals combining cultures willy-nilly? No, this is YouTube becoming Hollywood's valued ally, from tracking intellectual property to realizing the culture industries' dream: permitting corporations to engage in product placement each time their own copyright is infringed online, and learning more and more about their audiences. Such social networks are culture companies that rely on unpaid labour for their textuality and seek, at the core of their business models, to obfuscate distinctions in viewers' minds between commercials and programs via participatory video ads (Keen 2007).

It is often alleged that political economists of the media have not accounted for the creativity of audiences/consumers. But they are well aware of this capacity. In the 1950s, Dallas Smythe (1954) wrote that 'audience members act on the program content. They take it and mold it in the image of their individual needs and values.' He saw no necessary contradiction between this perspective and his other principal intellectual innovation, namely that audience attention – presumed or measured – was the commodity being sold in the commercial TV industry, by stations to advertisers. Similarly, in his classic 1960s text *Mass Communications and American Empire* (1992), Herb Schiller stressed the need to build on the creativity of audiences by offering them entertaining and informative media. And at the height of his work in revolutionary societies, from Latin America to Africa, Armand Mattelart (1980) recognized the relative autonomy of audiences and their capacity and desire to generate cultural meanings.

Media texts and institutions are not just signs to be read; they are not just coefficients of political and economic power; and they are not just innovations. Rather, they are all these things. Culture is indeed adaptive. It is a hybrid monster, coevally subject to text, power and science – all at once, but in contingent ways (Latour 1993). I therefore favour a tripartite approach to analysing it: a reconstruction of 'the diversity of older readings from their sparse and multiple traces' (157); a focus on 'the text itself, the object that conveys it, and the act that grasps it' (161–63); and an identification of 'the strategies by which authors and publishers tried to impose an orthodoxy or a prescribed reading on the text' (Chartier 1989, 166). Amongst the pieces I was sent for this collection, the essays by Anthony May, Bill Grantham, Albert Moran, and Anthony Fung and Mickey Lee stand out in this respect.

This materialist history must be evaluated inside consideration of the wider political economy. As Jacques Attali (2008) explains, lengthy historical cycles see political-economic power shift between cores. A new 'mercantile order forms wherever a creative class masters a key innovation from navigation to accounting or, in our own time, where services are most efficiently mass produced, thus generating enormous wealth'. Manuel Castells (2007) has coined the term 'mass self-communication' to capture this development, which sees affective investments by social movements and individuals matched by financial and policing investments by corporations and states. Papers here by John Sinclair and Rowan Wilken, and David Rowe and Callum Gilmour, undertake such work.

New eras in communication also index homologies and exchanges between militarism, colonialism and class control. The networked-computing era has solidified a unipolar world of almost absolute US dominance, with a share taken by other parts of the world economic triad in Japan and Western Europe. None of that has changed or been even mildly imperilled by cultural recombination or anything else. China and India provide many leading software engineers, but they lack domestic venture capitalists, military underpinnings to computing innovation and successful histories of global textual power at the mainstream level as per Sony, the BBC, Hollywood or the Pacific Northwest.

There *are* alternatives to the hypocrisy of neo-liberalism. Bruno Latour thinks that global interdependence generated by life in a risk society may be shifting us 'from a time of succession to a time of co-existence', where historicity and commonality prevail (Latour with Kastrissianakis 2007). That is where our recombinations will inevitably take form: in the complex sociality of collective experience and struggle.

Notes

1. The numbers on immigration invoked here can be found in Miller (2007).
2. The material in this section draws on Miller et al. (2005).

References

Andrejevic, M. 2007. *iSpy: Surveillance and power in the interactive era.* Lawrence: University Press of Kansas.

Attali, J. 2008. This is not America's final crisis. *New Perspectives Quarterly*, Spring: 31–3.

Castells, M. 2007. Communication, power and counter-power in the network society. *International Journal of Communication* 1: 238–66.

Chartier, R. 1989. Texts, printings, readings. In *The new cultural history*, ed. L. Hunt, 154–75. Berkeley: University of California Press.

Cohen, N. 2008. The valorization of surveillance: Towards a political economy of Facebook. *Democratic Communiqué* 22, no. 1: 5–22.

Collins, L. 2008. YouTube generation no match for the man. *Engineering & Technology*, 24 May– 6 June: 40–1.

Eagleton, T. 1982. The revolt of the reader. *New Literary History* 13, no. 3: 449–52.

Foucault, M. 2008. *The birth of biopolitics: Lectures at the Collège de France, 1978–79.* Trans. Graham Burchell. Ed. Michel Senellart. Houndmills: Palgrave Macmillan.

Grossman, L. 2006. *Time*'s Person of the Year: You. *Time*, 13 December. http://www.time.com/ time/magazine/article/0,9171,1569514,00.html (accessed 2 Jan 2009).

Hall, Stuart and P. Whannell. 1965. *The Popular Arts*. New York: Pantheon.

Jenkins, H., and J. Hartley. 2008. Is YouTube truly the future? *Sydney Morning Herald*, 25 June. http://www.smh.com.au/news/opinion/is-youtube-truly-the-future/2008/06/24/1214073239134.html (accessed 2 Jan 2009).

Keen, A. 2007. *The cult of the amateur: How today's Internet is killing our culture and assaulting our economy*. London: Nicholas Brealey.

Kiss, J. 2008. Now for something completely different. *Guardian*, 24 November. http://www.guardian.co.uk/media/2008/nov/24/googlethemedia-digitalmedia

Latour, B. 1993. *We have never been modern*. Trans. C. Porter. Cambridge, MA: Harvard University Press.

Latour, B. with K. Kastrissianakis. 2007. We are all reactionaries today. *Re-Public: Re-Imagining Democracy – English Version*, 22 March. http://www.re-public.gr (accessed 2 Jan 2009).

MacCabe, C. 2008. An interview with Stuart Hall, December 2007. *Critical Quarterly* 50, nos. 1–2: 12–42.

Mattelart, A. 1980. *Mass media, ideologies and the revolutionary movement*. Trans. M. Coad. Brighton: Harvester Press; Atlantic Highlands, NJ: Humanities Press.

McChesney, R.W. 2007. *Communication revolution: Critical junctures and the future of media*. New York: New Press.

Miller, T. 2007. Culture, dislocation, and citizenship. In *Global migration, social change, and cultural transformation*, ed. E. Elliott, J. Payne, and P. Ploesch, 166–86. New York: Palgrave Macmillan.

———. 2009. Can natural Luddites make things explode or travel faster? The new humanities, cultural policy studies, and creative industries. In *Media industries: History, theory, and method*, ed. Jennifer Holt and Alisa Perren, 184–98. Malden, MA: Wiley/Blackwell.

Miller, T., G. Nitin, J. McMurria, R. Maxwell, and T. Wang. 2005. *Global Hollywood 2*. London: British Film Institute.

Postrel, V. 1999. The pleasures of persuasion. *Wall Street Journal*, 2 August: 18.

Schiller, H.I. 1992. *Mass communications and American empire*. 2nd ed. Boulder, CO: Westview.

Smythe, D. 1954. Reality as presented by television. *Public Opinion Quarterly* 18, no. 2: 143–56.

Vincent, N. 2000. *Lear, Seinfeld*, and the dumbing down of the Academy. *Village Voice*, 2–8 February. http://www.villagevoice.com/2000-02-01/nyc-life/hop-on-pop (accessed 2 Jan 2009).

YouTube rolls out filtering tools. 2007. *BBC News*, 16 October. http://newsvote.bbc.co.uk (accessed 2 Jan 2009).

Index